THE KENNEDY SPACE CENTER STORY

1991 edition
Compiled and written by NASA Public Affairs, Kennedy Space Center, Fla.
Produced by Graphic House Inc., Orlando, Fla.

contents

the KENNEDY SPACE CENTER STORY

the KENNEDY SPACE CENTER STORY

An entire chapter of U.S. history has been written at the John F. Kennedy Space Center (KSC). As the departure site for our first journey to the Moon, and hundreds of scientific, commercial, and applications spacecraft, and now as the base for Space Shuttle launch and landing operations, KSC plays a pivotal role in the nation's space program.

Located on the east coast of Florida approximately midway between Jacksonville and Miami, the 140,000 acres (56,700 hectares) controlled by the Center represent a melding of technology and nature. Wildlife thrives here, alongside the immense steel-and-concrete structures of the nation's major launch base. KSC is a national wildlife refuge, and part of its

Wildlife and high technology exist side by side at the Kennedy Space Center. The same pair of Southern bald eagles--an endangered species--may have inhabited this nest for the past 25 years. The birds have an average life space of 30 years.

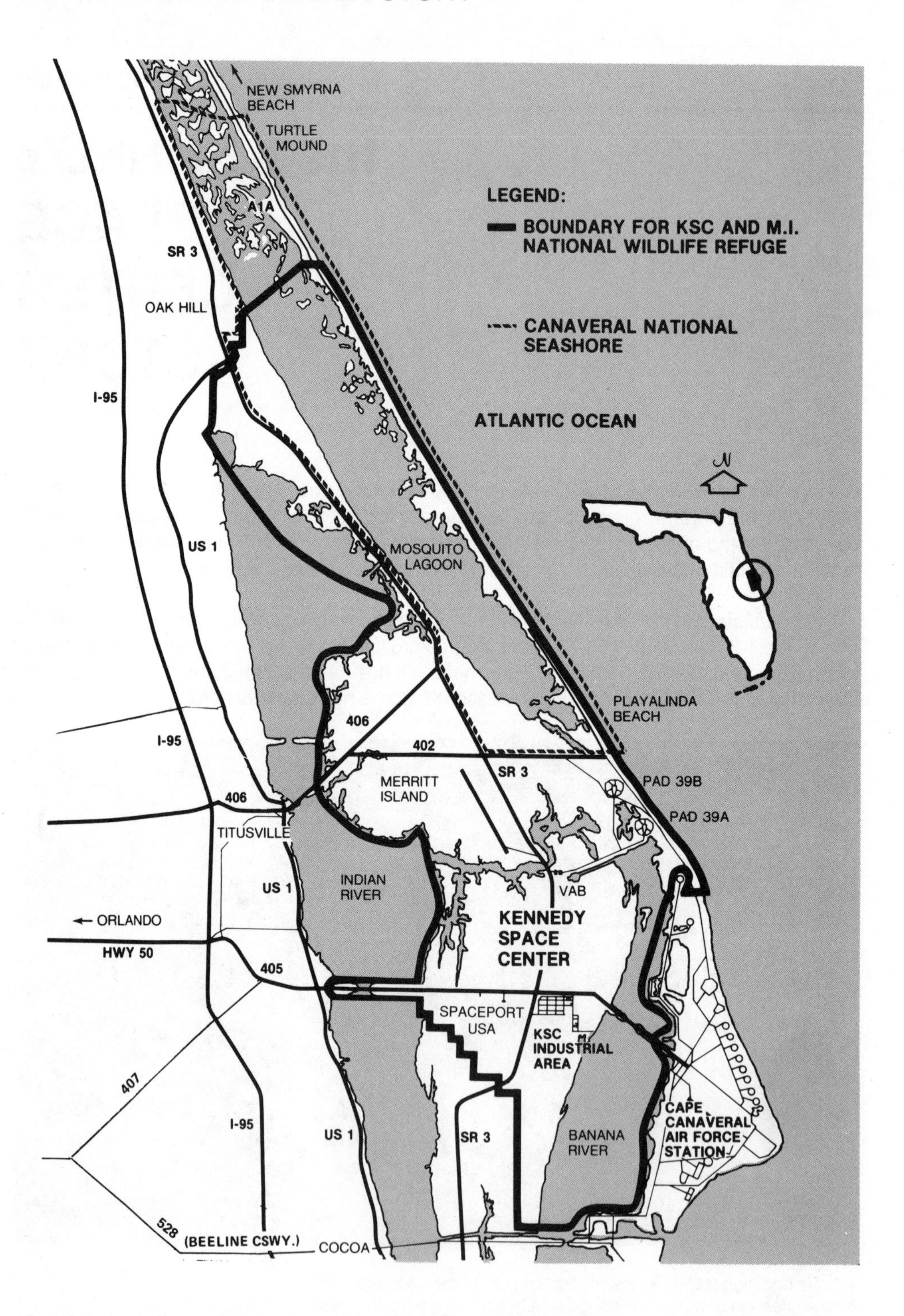

Map of Kennedy Space Center, Cape Canaveral Air Force Station, and adjacent population centers in the surrounding Florida east coast area.

coastal area is a national seashore by agreement between the National Aeronautics and Space Administration (NASA) and the Department of the Interior. More than 200 species of birds live here year-round, and in the colder months large flocks of migratory waterfowl arrive from the North and stay for the winter. Many species of endangered wildlife are native to this area: the Southern bald eagle, brown pelican, manatee, peregrine falcon, green sea turtle, and Kemp's Ridley sea turtle.

KSC extends about 34 miles (55 kilometers) from north to south and measures 10 miles (16 kilometers) at its widest point. Located primarily on Merritt Island, the facility is bounded on the east by the Atlantic Ocean and the Banana River, and on the west by the Indian River. The northern boundary is some 25 miles (40 kilometers) south of Daytona Beach, and the southern tip is just across the Banana River from Port Canaveral.

Essentially flat, KSC land averages about five feet (1.5 meters) above sea level. Extensive marshes and scrub vegetation, including saw palmetto, blanket most of the terrain. Cabbage palm, slash pine and oak grow on higher ground. Long rows of Australian pine protect citrus groves planted by early settlers on Merritt Island.

Archaeologists have uncovered burial mounds and shell middens (refuse piles) left by small bands of prehistoric Indians who inhabited the area thousands of years ago. These Indians were attracted by the abundance of marine food found in the marshes and saltwater creeks in the area.

Spanish fleets en route from the New World to the mother country once sailed the Gulf Stream off Cape Canaveral. Treasure hunters still search for traces of galleons which foundered off the coast and deposited their contents on the ocean floor.

There are more than 1,500 acres (607 hectares) of citrus groves on the Center. These lands are leased to individuals—in many cases the original owners—who care for the trees and harvest their fruit. Beekeepers collect honey from and maintain the hives of bees essential to the pollination of citrus trees. The lease arrangements are administered by the Merritt Island National Wildlife Refuge.

The nerve center of KSC is Launch Complex 39. This is the location of the Vehicle Assembly Building, where Saturn V vehicles were once prepared for launch. This massive building is now the NASA assembly site for the Space Shuttle.

Some 3.5 miles (5.6 kilometers) to the east of the assembly building are the two launch pads where journeys into space begin. Five miles (eight kilometers) south is the KSC Industrial Area, where many of the Center's support facilities are located. First Apollo and now Shuttle crews prepare here for the next mission. Here also are the administrative headquarters for KSC operations, the offices of the Center director and other NASA and contractor managers.

Spaceport USA, the KSC visitors center, is located on the NASA Causeway (an extension of State Road 405), south of Titusville, and six miles east (9.6 kilometers) from U.S. Highway 1. Available to visitors at no cost are displays of spacecraft, rockets and space equipment; space and aeronautic exhibits; and space science films and demonstrations. The IMAX, one of two theatres, has an admission fee. The IMAX production is an experience much like actually being there during a Shuttle liftoff, or working with astronauts in the vast openness of space. Also available for a modest fee are conducted bus tours through Kennedy Space

A marine biologist releases tagged green sea turtles on the beach behind Launch Complex 39 at the Kennedy Space Center. This turtle is one of the many species of endangered wildlife indigenous to the Cape Canaveral area.

An American alligator, once a member of a threatened species, thrives with its neighbors in the protected waters of canals dug to drain the Space Shuttle runway.

Center and adjacent Cape Canaveral Air Force Station.

As the role of the spaceport changed with the demands of the national space program, the organization of KSC altered to meet those needs. In keeping with NASA's philosophy of using private industry and the nation's universities wherever possible, the majority of KSC employees work for aerospace contractors. In addition to the work tasks required to assemble, process and launch the Space Shuttle, its payloads and crews, a variety of support functions are necessary to keep this large installation operating. These include day-to-day supply, transportation, grounds maintenance, documentation, drafting, and design engineering. Contractors bid competitively on these functions, and are awarded contracts based on their bids. Contracts are administered by KSC's NASA civil service work force. (In keeping with the purposes of this history, a company is referred to by its name at the time of contract performance. In some instances, these names have since changed).

KSC is one of 12 NASA field installations spread across the nation. NASA Headquarters, Washington, D.C., formulates policy for the agency and coordinates the specialized activities of each NASA facility. The other NASA installations are:

- Ames Research Center, Moffett Field, Calif.
- Hugh L. Dryden Flight Research Facility, Edwards, Calif.
- Goddard Space Flight Center, Greenbelt, Md.
- Jet Propulsion Laboratory, Pasadena, Calif.
- Lyndon B. Johnson Space Center, Houston, Texas.
- Langley Research Center, Hampton, Va.
- Lewis Research Center, Cleveland, Ohio.
- George C. Marshall Space Flight Center, Huntsville, Ala
- Michoud Assembly Facility, New Orleans, La.
- John C. Stennis Space Center, Bay St. Louis, Miss.
- Wallops Flight Facility, Wallops Island, Va.

The technology, equipment and concepts developed during the early days of space exploration were the building blocks for an operational spaceport. We have broken the bonds of Earth and traveled into space and back on many occasions. Mercury, Gemini, Apollo ... America's space travelers have given new meaning to these names from Greek and Roman mythology. In this era of the Space Shuttle, the journey into space continues. These pages tell of the people and the programs responsible for creating a gateway to the solar system, and beyond.

Directors of the Kennedy Space Center

Dr. Kurt H. Debus
1962-1974

Lee R. Scherer
1975-1979

Richard G. Smith
1979-1986

Lt. Gen. Forrest S. McCartney
(USAF, retired)
1986-1991

Capt. Robert L. Crippen
(USN, retired)
1992-

chapter 1

ORIGINS

With the notable exception of Dr. Robert H. Goddard's pioneering work with liquid propellant rockets in the 1920s and 1930s, American interest in rocketry and space exploration prior to World War II was restricted to amateur rocket clubs and the fertile outpourings of science fiction writers. With the outbreak of war, military demands led to the development of a host of rocket-powered battlefield weapons for use against tanks, armored vehicles and submarines and also for barrages in support of troop landings and advancements. Space science, a loose term at the time, was limited primarily to weather and upper atmospheric studies using balloons and small sounding rockets. Although the idea of putting artificial satellites into orbit around the Earth for military and scientific purposes had been explored by the armed forces and various civilian agencies, it never passed the talking stage; rocketry had not yet progressed to the point where such far-out schemes were feasible.

Toward the end of the war, American interest in rocket technology had increased dramatically. This was mainly because of the impact of the successful German V-2 rocket development on American military and scientific circles. Military planners saw the long-range V-2 as the shape of things to come in the dawning nuclear age. Scientists viewed it as a tool for high-altitude research and the forerunner of larger rocket systems for the exploration of space. Eager to cash in on this technological bonanza, the U.S. Army brought a number of German rocket experts and almost 100 confiscated V-2 rockets to this country following the end of hostilities.

The Army began testing V-2s in 1946 at its Ordnance Proving Grounds at White Sands, N.M. Here, German scientists and technicians, headed by Dr. Wernher von Braun, developer of the V-2, worked alongside their American counterparts in putting reassembled V-2s to use for research. In the course of the next five years, teams from each of the three armed services, aerospace industries and universities—partners in America's missile and space development—assembled information from the successful launchings of 40 instrumented V-2s. While the tests yielded invaluable data in high-altitude research, the emphasis—and congressional appropriations—were tuned to the development of intermediate and intercontinental range ballistic missiles for national defense. As the range and sophistication of the V-2 and follow-on rocket systems increased, it became evident that a new, long-range test site was needed. In October 1949, President Harry S. Truman established the Joint Long Range Proving Grounds at Cape Canaveral, Fla.

The Cape was ideal for testing missiles. Virtually undeveloped, it enabled personnel to inspect, fuel and launch missiles without danger to nearby communities. The area's climate also permitted year-round operations, and rockets could be launched over water

A captured German V-2 missile was destined to become the forerunner of rockets that would peacefully exploit space for the benefit of people on Earth. Bumper, a V-2 with an Army WAC (Without Any Control) Corporal second stage, was the first missile launched from Cape Canaveral—on July 24, 1950.

instead of populated areas. A chain of islands extending southeastward from Grand Bahama to Ascension provided sites for tracking stations to follow the progress of missiles in flight. Some years later, many of these same factors led to selection of the area adjacent to Cape Canaveral—Merritt Island—as the location of the Kennedy Space Center.

After the proving grounds were established, the Air Force took over the nearby Banana River Naval Air Station—renamed Patrick Air Force Base in honor of Maj. Gen. Mason Patrick, the first chief of the Army Air Corps—located 20 miles (32 kilometers) south of the Cape. Here, in 1951, the Air Force established headquarters for the Air Force Missile Test Center, which included a range. These facilities were redesignated in 1964 as the Air Force Eastern Test Range, which became Detachment 1, Space and Missile Test Center in 1977, and was redesignated the Eastern Space and Missile Center in 1979.

An Army team from White Sands conducted the first rocket launch from Cape Canaveral on July 24, 1950. The rocket was called Bumper 8, a modified V-2 with a WAC (Without Any Control) Corporal stage mounted on top. It achieved an altitude of

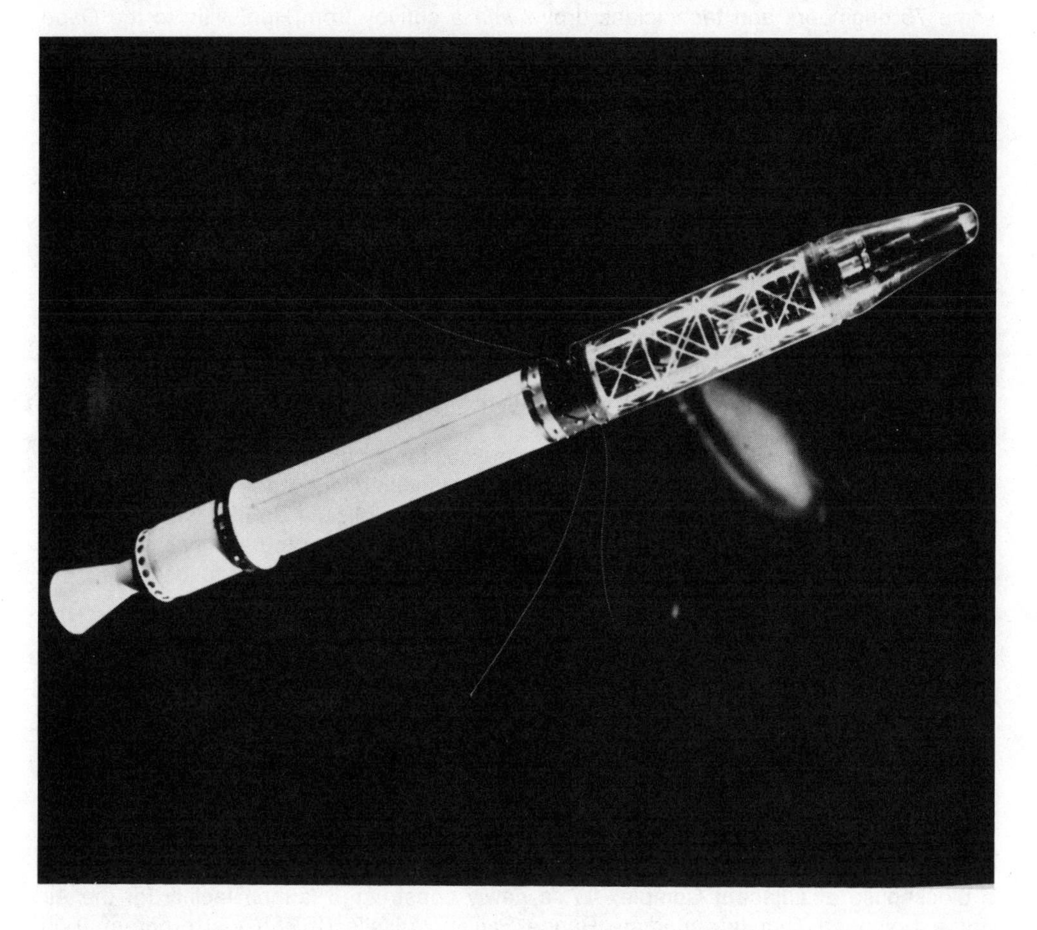

Explorer 1, launched Jan. 31, 1958, by a U.S. Army team directed by Dr. Kurt Debus, helped bolster sagging U.S. prestige.

10 miles (16 kilometers). The launch facilities used for Bumper 8 contrast sharply with those in use on the Cape today. For that primitive launch, Army technicians employed a painter's scaffold as a gantry to service the rocket before launch, and the control center was a converted tarpaper bathhouse surrounded by sandbags.

Shortly afterwards, the Army decided to consolidate its rocket and guided missile programs at the Redstone Arsenal, a former chemical development center near Huntsville, Ala. The first projects assigned to the new Ordnance Guided Missile Center, later renamed the Army Ballistic Missile Agency, were development of the Redstone and the Jupiter missiles. The Redstone was a direct descendant of the V-2, which would have a 200-mile (322-kilometer) range, and the Jupiter was a more powerful version of the Redstone with a 1,750-mile (2,800-kilometer) range. The missile agency was commanded by Maj. Gen. J. B. Medaris. Von Braun headed the agency's Development Operations Division and Dr. Kurt H. Debus, a von Braun associate, was in charge of the Missile Firing Laboratory, which was responsible for launch operations.

The Army began testing Redstone missiles at the Cape in August 1953. Debus and some 75 engineers and technicians drove with a convoy from Huntsville to the Cape for each test. There, they assembled and launched a missile, then returned to Alabama to work on the next missile on the assembly line. When the Army later contracted with Chrysler Corp. to produce Redstones and Jupiters in Michigan, the launch team moved permanently to the Cape to conduct flight testing and assist in training U.S. and foreign military crews in handling these weapons.

By the mid-1950s, rocket technology in the United States had reached the stage where serious consideration was being given to proposals to launch Earth satellites. The opportunity surfaced in July 1955 when President Dwight D. Eisenhower announced that the United States would place a satellite into orbit as part of its contribution to the 1957-58 International Geophysical Year. The Soviet Union also announced its intention of orbiting a satellite during that year, even suggesting that its artificial moon would be much bigger than any the United States might attempt to launch.

The competition among the U.S. armed services for the honor of launching America's first satellite was lively. The Army proposed using a modified four-stage Redstone vehicle. The Air Force pushed its Atlas Intercontinental Ballistic Missile (ICBM) which was still under development, and the Naval Research Laboratory promoted an improved three-stage version of its workhorse Viking high-altitude research rocket, to be called the Vanguard.

Project Vanguard was selected for America's satellite program because it provided an opportunity to develop a new rocket system for civilian rather than military purposes. Furthermore, officials pointed out, Project Vanguard was a scientific program and, unlike the Army and Air Force proposals, would have little or no impact on the nation's critical task of developing ballistic missiles for national defense.

The Navy's Vanguard operations group arrived at the Cape in late 1955 and began preparing for launch operations. They were assigned Pad 18A for launchings, and shared a blockhouse at adjacent Complex 17, a newly constructed launch facility for the Air Force's planned Thor Intermediate Range Ballistic Missile (IRBM) test program. Two Vanguard test vehicles were launched successfully on suborbital flights in late 1956 and 1957. Then came Sputnik.

Russia's launch of the world's first artificial satellite on Oct. 4, 1957, had a profound effect on the American people and the governmental agencies involved with satellite development. Moreover, the size and weight of Sputnik 1—184 pounds (83 kilograms)—made this country acutely aware that the Soviets had developed rockets far more powerful than any in the American arsenal. This point was driven home even more dramatically a month later when the Soviets launched Sputnik 2, a 1,120-pound (508-kilogram) satellite carrying a dog named Laika. Although there was no way the United States could hope to match the Russian feats, some prestige might be regained by orbiting an American satellite.

The pressure, along with the eyes of the world, was clearly on the Vanguard launch team. The pressure intensified when the U.S. secretary of defense directed the Army launch team to prepare for a satellite launching in the event the Vanguard failed to achieve its objective. Following a third successful test flight, the Vanguard team attempted a satellite launch on Dec. 6, 1957, a little over two months after Sputnik 1. A few seconds after liftoff, when Vanguard had risen about four feet (1.2 meters), its engine lost thrust and the rocket and its beeping payload fell back upon the launch pad and exploded in a flaming fireball that further seared American confidence. Before the Vanguard team could ready another vehicle for flight, the Army team directed by Dr. Debus launched its 30-pound (13.6-kilogram) Explorer 1 satellite into orbit on Jan. 31, 1958, from nearby Complex 26A, using a modified Jupiter-C called Juno 1. America was in space. Explorer 1, built for the Army by the Jet Propulsion Laboratory of the California Institute of Technology, made up in quality what it lacked in size. An onboard experiment designed by Dr. James Van Allen of the State University of Iowa (now the University of Iowa) detected the Earth's radiation belt, subsequently named the Van Allen Radiation Belt.

The grapefruit-size Vanguard became America's second satellite—and the world's first spacecraft powered by solar cells.

The Army attempted a second satellite launch in March, but it failed when the Juno 1's fourth stage did not ignite. On March 17, 1958, Vanguard placed America's second satellite into orbit, a 3.5-pound (1.6-kilogram) sphere that will stay aloft for 2,000 years. Called Vanguard 1, the tiny satellite carried the first solar cells into space and is now the oldest and smallest spacecraft in orbit. Although the Army team remained America's first heroes of the space age, Project Vanguard personnel had the satisfaction of knowing

A modified Jupiter-C booster, called Juno 1, became the first launch vehicle to place an American satellite—Explorer 1—in Earth orbit.

that in record time—only two years, six months and eight days—they had developed a new high-performance three-stage launch vehicle, a launching facility, a worldwide tracking system and a range instrumentation network. More importantly, they had accomplished their mission: placing a satellite into orbit during the International Geophysical Year.

Despite America's success with Explorer and Vanguard, the Russian Sputniks engendered a widespread clamor that the United States embark upon a vastly expanded space program. NASA came into being on Oct. 1, 1958, absorbing some 8,000 personnel and the laboratories of the 43-year-old National Advisory Committee for Aeronautics. The transfer to NASA included Wallops Station in Virginia and four research centers: Langley Memorial Aeronautical Laboratory (renamed Langley Research Center), Hampton, Va.; Lewis Flight Propulsion Laboratory (renamed Lewis Research Center), Cleveland, Ohio; Ames Aeronautical Laboratory (renamed Ames Research Center), Moffett Field, Calif.; and the High Speed Flight Station (renamed the Flight Research Center and later the Dryden Flight Research Facility), Edwards, Calif.

NASA also absorbed about 150 Project Vanguard personnel who, along with other elements from the scientific and military communities, formed the nucleus of the Beltsville Space Center (renamed the Goddard Space Flight Center) at Greenbelt, Md., established in January 1959. The Vanguard Operations Group at the Cape, renamed the Goddard Space Flight Center's Field Projects Branch, officially became NASA's first launch team. Under the direction of Robert H. Gray, who had served as test conductor for the Vanguard missions, the Goddard team was responsible for launching the majority of the nation's pioneering satellite programs, including lunar and planetary probes and the world's first weather and communications satellites.

During this period, the Cape was transformed from scrubland into a major launch base. The Army continued testing its Redstones and Jupiters and a new short-range field missile called the Pershing. The Navy was busy developing its submarine-launched Polaris missile system. The Air Force pushed ahead with its Thor IRBM and its Atlas, Titan and Minuteman ICBM programs. The Thor later became the booster stage for NASA's dependable Delta launch vehicle, which has placed more satellites into orbit than any other rocket in the nation's fold. The Atlas and the Titan also were destined for NASA duty, as booster stages for the agency's manned and unmanned space programs. Launch complexes to support military launch operations sprang up the length and width of the Cape. Many were modified later to support NASA space activities.

Meanwhile, at the Army Ballistic Missile agency in Huntsville, the von Braun team was designing and developing a super booster. Under a study financed by the Advanced Research Projects Agency of the Department of Defense, the team had clustered eight rocket engines to see if a single stage could produce 1,000,000 pounds (4,500,000 newtons) of thrust, far more than any rocket known. However, in August 1959, the Defense Department decided it had no need for a rocket of this size and suggested that it might serve as a booster in NASA's space program. A month later, the program was transferred to NASA.

In December 1959, an agreement was signed by the Department of Defense and NASA, transferring the von Braun team of 5,000 civil servants, including the Missile Firing Laboratory directed by Debus, from the Army to NASA. The transfer involved $100 million

Cape Canaveral, a scrub-covered real estate development that never quite happened, was transformed into a major launch base for rocket development.

in laboratories, test stands, equipment, shops and office buildings. Von Braun was appointed director of the newly established George C. Marshall Space Flight Center in Huntsville and assigned the task of developing heavy space launch vehicles. Debus was appointed director of Marshall's Launch Operations Directorate at Cape Canaveral, the organization that would form the nucleus of the Kennedy Space Center some 3 1/2 years later.

While NASA was in its formative stages, the Soviet Union progressed with its own space program. On Sept. 12, 1959, Luna 2 impacted on the Moon, and less than a month later Luna 3 took the first pictures of the dark side of the Moon. In August 1960, the Soviets recovered a spacecraft which had carried animals into Earth orbit. In February 1961, they launched a space probe which obtained measurements of the environment of Venus. Then, on April 12, 1961, Russian cosmonaut Yuri Gagarin became the first person to travel in space. The U.S. followed by launching Alan Shepard on his historical suborbital flight less than a month later.

Shepard's flight, the nation's initial manned space flight effort, narrowed the space gap between the United States and Russia. But even more ambitious undertakings were planned by the United States. In May 1961, President John F. Kennedy fired the public imagination by announcing that the United States would fly men to the Moon and back

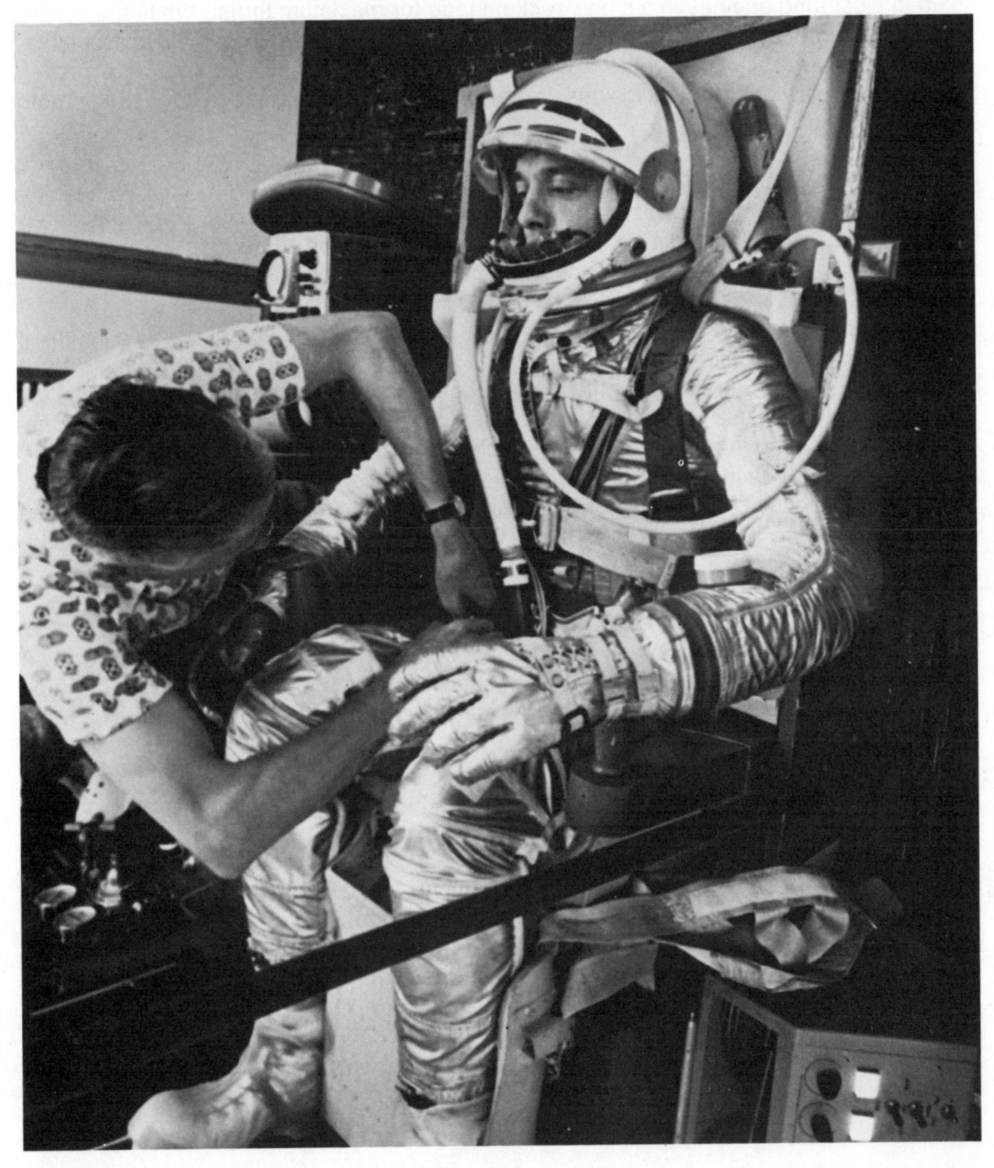

The "space gap" narrowed with the historic suborbital space flight of Alan Shepard, less than a month after Russia's first manned space flight.

within the decade. His challenge elicited congressional support for a program which required rockets far more powerful than any then available, and spacecraft designed to protect men from the hostile environment of space during the 500,000-mile (800,000-kilometer) journey to and from the Moon.

The program was Apollo, and the vehicle that would launch the Apollo spacecraft and its three-man crew to the Moon was the Saturn V. This three-stage rocket would generate about five times the thrust of the Saturn I, just then reaching its flight phase. The Saturn series of vehicles were the result of experiments carried out by the von Braun team in clustering engines in a single rocket stage for maximum thrust. While the Saturn V was taking shape on the drawing boards, a suitable location had to be found to assemble, service and launch it. Although the Cape's 17,000 acres (6,885 hectares) had proven adequate for previous space missions, larger facilities would be needed for the mammoth Moon rocket.

Dr. Debus, representing NASA, and Lt. Gen. Leighton I. Davis, representing the Department of Defense, organized a joint study to find a new launch site. They considered Hawaii, Texas, the California coast, an island off the coast of Georgia, islands in the Caribbean, and Merritt Island (adjacent to the Cape) as possible sites.

The study concluded that Merritt Island offered compelling advantages. Several small communities were within easy driving range, and larger cities like Daytona Beach, Vero Beach, and Orlando were only slightly further. Locating on Merritt Island also would allow NASA to share facilities of the Atlantic Missile Range, avoiding costly duplication. Only at this location could the same NASA launch organization continue operations on the Cape Canaveral complex while building the spaceport. Debus and General Davis recommended the acquisition of the northern part of Merritt Island. The choice was endorsed by NASA and the Defense Department. Congress authorized NASA to acquire the property.

The space agency began acquisition in 1962, taking title to 83,894 acres (33,952 hectares) by outright purchase. It negotiated with the state of Florida for use of an additional 55,805 acres (22,600 hectares) of state-owned submerged land, most of which lies within the Mosquito Lagoon. The investment in property reached approximately $71,872,000.

In July 1962, the Launch Operations Directorate at the Cape was separated from the Marshall Space Flight Center by executive order. It became the Launch Operations Center, an independent NASA installation, with Debus as its first director. It was renamed the John F. Kennedy Space Center in December 1963, in honor of America's slain president. In December 1964, launch elements of Houston's Manned Spacecraft Center (now the Johnson Space Center) were transferred to the Kennedy Space Center. The following October, Goddard Space Flight Center's Field Projects Branch on the Cape was incorporated into the Kennedy Space Center.

The challenge had been issued and accepted. Next came the task of meeting that challenge through the design, construction and operation of a complete spaceport.

chapter 2

EXPENDABLE LAUNCH VEHICLE OPERATIONS

While development of the lunar program moved forward, unmanned launches continued unabated from Cape Canaveral. Part of the original Vanguard Naval Research Laboratory team became the Launch Operations Branch of the Goddard Space Flight Center after NASA was established in 1958. This team became a part of the Kennedy Space Center in October 1965.

The team completed its planned series of Vanguard launches while NASA was developing more powerful vehicles. Because they could be used only once and their components were not recoverable, these unmanned rockets also were referred to as expendable launch vehicles (ELVs). In April 1959, NASA awarded a contract to the

Early launch vehicles used by NASA: from left are Thor-Agena, Juno, and Atlas-Agena.

Douglas Aircraft Co. for the design, fabrication, testing and launch of an improved version of the three-stage Thor-Able expendable booster, to be called Thor-Delta (later simply the Delta). The first stage was a Thor Intermediate Range Ballistic Missile and the second and third stages were modifications of the second and third stages of the Vanguard. The Goddard group was given the responsibility for supervising the checkout and launch of Delta vehicles. They were still performing these tasks when phased over into KSC.

The Delta became the "workhorse" of NASA's ELV family, undergoing a number of upgrades in power until it could place 2,800 pounds (1,270 kilograms) in geosynchronous transfer orbit, more than 20 times its original payload capability. Of the 182 Delta launches NASA conducted through 1988, 170 were successes.

While the Cape was the primary Delta launch site, many were also launched from the West Coast. NASA used launch pads at Vandenberg Air Force Base in California to achieve polar or other north-south orbits required for certain meteorological and Earth resources satellites, as well as cosmic explorer spacecraft.

The second vehicle added to the NASA unmanned medium launch vehicle category was the Thor-Agena. The Agena was a powerful upper stage developed for the Air Force by the Lockheed Propulsion Co. It used liquid propellants and had inflight shutdown and restart capabilities. A test flight on Jan. 15, 1962, achieved most of its objectives. After two test flights, a Thor-Agena placed the huge Echo 2 balloon into orbit on Jan. 25, 1964. This spherical balloon, 135 feet (41.1 meters) in diameter, remained in orbit for two years and was the object of numerous early communications experiments by scientists from the United States, the United Kingdom, and the Soviet Union. The Thor-Agena remained in service until April 1970, with a total of 12 operational missions. This was the first NASA vehicle to be launched from Vandenberg Air Force Base. It was also the first NASA vehicle to have solid propellant rockets strapped to its first stage for additional thrust—a technique that became standard with the Delta.

Almost concurrently with the Thor-Agena, NASA developed the Atlas-Agena, a much more powerful vehicle. The Atlas stage had been developed as an Intercontinental Ballistic Missile by General Dynamics/Convair for the Air Force. When mated with the Agena, the vehicle had the capability of placing spacecraft in lunar or interplanetary trajectories. The first operational launch was on Jan. 30, 1964, a Ranger mission to impact on the Moon and take photographs during the descent to the surface. The vehicle performed well, but the Ranger camera system failed. The second Ranger mission was a success, however, returning the first close-up photographs of the lunar surface. There were 19 Atlas-Agena missions in all—including four Rangers to the Moon; five Lunar Orbiters, the first spacecraft from the United States to enter orbit around and photograph another planetary body; the first Mariner spacecraft sent to Venus and Mars; the first three Applications Technology Satellites in Earth orbit; the first Orbiting Astronomical Observatory; and three Orbiting Geophysical Observatories. The last NASA Atlas-Agena was launched on March 4, 1968.

The Air Force also continued work on vehicle development programs, though its aims and capabilities were dissimilar to those of NASA. One early project was a hydrogen-burning stage called Centaur, that would be far more powerful than those using less-volatile kerosene fuel. The service had inherited this project as an engine development effort from the National Advisory Committee for Aeronautics, NASA's predecessor. On

July 1, 1959, the Air Force transferred the Centaur development program to NASA, in effect returning it to its originators. Marshall Space Flight Center was assigned management responsibility initially. The project was transferred to the Lewis Research Center between the first and second launches.

The Atlas-Centaur required new facilities, and Launch Complex 36, with two pads, was built on Cape Canaveral to accommodate it. The Atlas-Centaur development program was one of the most difficult in NASA history. The first launch on May 8, 1962, was a failure. Three of the next five missions, although trouble-plagued, were successes; one was a failure and the other only partially successful. After each flight, the analysis of system failures contributed to an overall understanding of vehicle performance. Engineering modifications and changes in the checkout procedures were instituted to correct the problems. Atlas-Centaur 8 failed in that the Centaur engines did not ignite for a second burn after a coast period of several minutes in low Earth orbit. But the design engineers and launch operations people were so certain they could correct the problem that the next flight, Atlas-Centaur 10, was scheduled for an important mission. On May 30, 1966, this vehicle carried the first Surveyor spacecraft to a spectacular accomplishment—the first soft landing of an American spacecraft on the Moon.

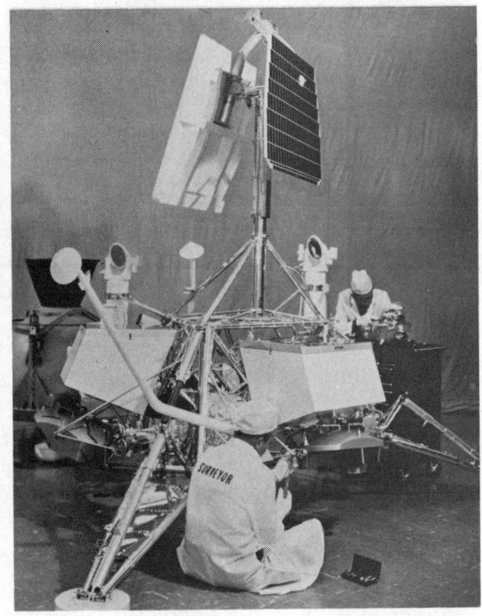

Atlas-Centaur rockets, left, have launched many satellites and space probes from Cape Canaveral, such as the Surveyor spacecraft, right, which soft-landed on the lunar surface.

After that tremendous success, the overall record of the new vehicle became quite good. Atlas-Centaur achieved seven successful launches of the Surveyors, of which five went on to land safely on the Moon. The vehicle sent Mariner spacecraft to Venus, Mercury and Mars, and Pioneer spacecraft to Jupiter and Saturn—feats that incredibly enriched

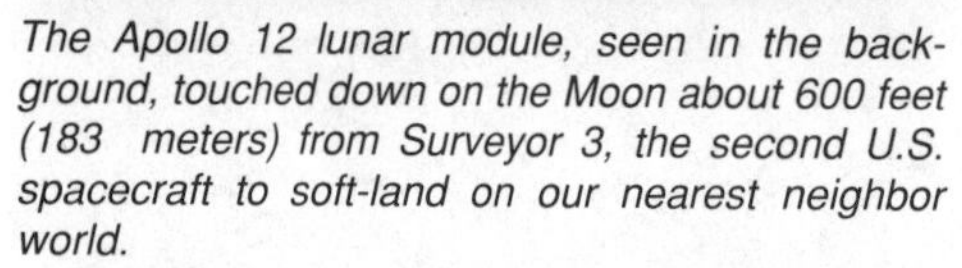
The Apollo 12 lunar module, seen in the background, touched down on the Moon about 600 feet (183 meters) from Surveyor 3, the second U.S. spacecraft to soft-land on our nearest neighbor world.

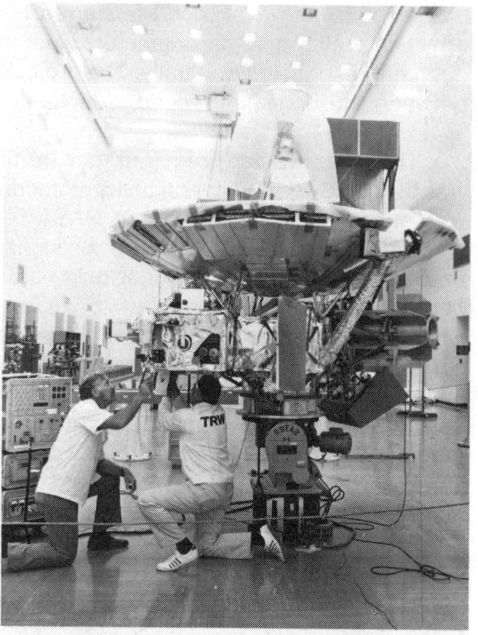

Pioneer 11, the first spacecraft to explore the ringed planet Saturn, was launched by KSC's Atlas-Centaur launch team and is now heading out of our solar system.

our understanding of the solar system. It also carried into Earth orbit many spacecraft too heavy for the other available vehicles. These included the very heavy Orbiting Astronomical Observatories, INTELSAT communications satellites, the larger Applications Technology Satellites, and many more.

In 1970, NASA planners foresaw a need for an unmanned launch vehicle with greater capability than the Delta or Atlas-Centaur. Several planned future missions—specifically the Viking automated laboratories to explore the Martian surface and atmosphere and two Voyagers to observe Jupiter and Saturn—would require a spacecraft too heavy for the Delta or Atlas-Centaur to carry. At that time the Space Shuttle was only a future possibility for NASA. After studying various alternatives, NASA decided that the fastest, most economical way to obtain the new heavy lift vehicle required would be to combine two existing systems. The planners decided to replace the small third stage on the Air Force's Titan IIIC vehicle with the far more powerful Centaur upper stage of the Atlas-Centaur. The new combination, called the Titan-Centaur, became the most powerful vehicle available in the United States' unmanned space program at that time.

In addition to the Vikings and Voyagers, Titan-Centaur launched two Helios spacecraft to study the Sun.

The launches of the two Voyager spacecraft to the outer planets in 1977 were the final assignments for Titan-Centaurs. Launch Complex 41, an Air Force launch site on Cape Canaveral, was borrowed by NASA from 1974 until 1977 for Titan-Centaur launches. Martin Marietta was the contractor for the Titan stages, and General Dynamics/Convair for the Centaur. Complex 41 later became the launch site for the most powerful unmanned U.S. rocket, the Titan IV, developed by Martin Marietta for the Air Force.

Launch of an expendable rocket did not carry the risk associated with placing humans in orbit. Nevertheless, it was a complex task involving many people and costing tens of millions of dollars. Integration and checkout of a NASA unmanned booster and its payload required that launch directorate personnel work closely with the spacecraft designers to prepare needed ground support systems. This was a process that began not just days or months, but sometimes years ahead of the actual launch date.

The careful advance preparation continued when the spacecraft was delivered to Cape Canaveral well ahead of the scheduled liftoff date, so it could be assembled and checked out by the manufacturer or owner. On scientific spacecraft, the scientists responsible for individual instruments and experiments often participated in the spacecraft checkout and launch activities.

Expendable launch operations personnel also had to determine radar and photographic requirements. Support provided by the Eastern and Western Test Ranges, operated by the U.S. Air Force, had to be coordinated.

Companies which built the launch vehicles were an integral part of the ELV launch team. The test conductor was always a contractor employee, while the launch director was always from NASA. Both NASA and contractor engineering staffs were on hand to resolve technical problems. Overall direction came from the launch director, while the test conductor manned the key console and provided detailed instructions to the launch team.

The ELV program also involved a close working relationship between KSC and other NASA centers. One of these was the Goddard Space Flight Center in Greenbelt, Md., which oversaw the design and development of the

The reliable and versatile Delta launch vehicle, using various combinations of strap-on boosters, was launched from both KSC and Vandenberg Air Force Base in California.

Delta for NASA. Goddard also supervised the design and packaging of many scientific and technological satellites, and continues to perform this activity today. The Lewis Research Center in Cleveland oversaw the design and development of the Atlas family of vehicles, as well as the Centaur upper stage.

The Jet Propulsion Laboratory at Pasadena, Calif., was and still is the control facility for the Deep Space Network used for tracking planetary exploration spacecraft such as the Voyagers, Vikings and Mariners. The Ames Research Center near San Francisco also plays an important part in planetary investigations. This has included designing and developing the Pioneer spacecraft which returned the first detailed information on the planets Venus, Jupiter and Saturn. The Langley Research Center in Hampton, Va., managed the design and construction of the Viking Landers, two spacecraft that sent back fantastic imagery and physical measurements from the surface of Mars, and the Lunar Orbiters whose photography of the Moon paved the way for the Apollo astronauts.

The original experienced launch team KSC inherited from Vanguard has gained and lost personnel through the years, and policy surrounding NASA expendable launch vehicle operations has undergone periodic redefinition. One noticeable change that occurred was the emergence of a large number of "reimbursable" launches—those undertaken for commercial customers, foreign governments or agencies, and other branches of the U.S. government.

These "customers" would usually buy or supply their own spacecraft, purchase the launch vehicle and service from NASA, and pay the expenses associated with launching their payload. Although NASA did not make a profit from these services, the numbers involved enabled the production lines to operate at a relatively high rate, resulting in mass production efficiencies which reduced the

The need for a launch vehicle more powerful than either Delta or Atlas-Centaur resulted in the Titan-Centaur, which sent the Viking and Voyager spacecraft on their exploratory quests toward other planets.

cost of each stage. Those which NASA purchased for its own use were then less expensive than they would be otherwise.

The owner assumed control of the satellite after it was placed in orbit. Most of the reimbursable missions involved applications satellites, derived from initial scientific investigations performed by NASA. The percentage of such missions peaked in 1980, when 83 percent of the spacecraft were launched on a reimbursable basis.

Manned space flight also had an impact on the unmanned payloads field. In January 1979, following a management reorganization, the unmanned launch operations directorate was assigned the added responsibility of processing payloads for the then nascent Space Shuttle program.

With the advent of the manned Shuttle, NASA envisioned that reliance on expendable launch vehicles would decline. The Shuttle concept offered a unique advantage over expendable launch vehicles—it was a reusable resource.

The first Shuttle flight occurred in April 1981. As one successful mission followed another, NASA implemented a corresponding phase-down in ELV launches.

An alternate role for expendable launch vehicles began to emerge in the early 1980s. In January 1983, the administration of President Ronald Reagan announced that the federal government would encourage private industry to build and operate ELVs to deliver commercial payloads—such as communications satellites—into orbit. Government facilities, including those at Cape Canaveral and Vandenberg, would be made available to private industry on a cost-reimbursable basis. The military also now regarded unmanned rockets as an invaluable backup to the Shuttle.

Further influencing government policy regarding ELVs was the Space Shuttle Challenger accident in January 1986. Six months later, NASA gave the commercial ELV industry a needed boost when the agency announced that commercial payloads such as communications satellites would no longer be deployed from the Shuttle.

About the same time, the Air Force announced plans to use unmanned vehicles for many of its future payloads. These decisions opened the door wide, not only to the big three American ELV manufacturers—General Dynamics, builder of the Atlas family of vehicles, Martin Marietta, the Titan, and McDonnell Douglas, the Delta—but other entrepreneurial firms eager to join the space race. These firms would not have to compete against the Space Shuttle to launch commercial payloads, although formidable competition was emerging from established or developing foreign firms and agencies.

NASA reassessed its own policy toward unmanned expendable rockets. The agency concluded that a mixed fleet of launch vehicles, rather than reliance on a single system, the Shuttle, was the best approach. In 1987, NASA announced its mixed fleet plan. "Expendable vehicles will help assure access to space, add flexibility to the space program, and free the Shuttle for manned scientific, Shuttle-unique, and important national security missions," NASA Administrator Dr. James Fletcher said.

The plan calls for procuring ELV launch service competitively whenever possible, except for a transitional first phase covering launches through 1991. During this transition,

NASA will procure ELV launch service non-competitively to address a backlog of space science missions. The agency's role will be the same throughout both phases, however: oversight of vehicle manufacture, preparation and launch, rather than direct management of it, as was the case in the past.

Under the transitional first phase, NASA will go through either the Air Force or directly to the vehicle manufacturer to obtain the best match between vehicle and payload for the time frame in which each will be needed. Vehicles being produced under contracts for the military, such as the Titan II, Titan IV, Delta II, Atlas E and Atlas II, will be procured via the Air Force.

If NASA should buy a commercially available vehicle, the manufacturer will provide not only the vehicle, but launch service as well. Falling into this category are several variants of expendable rockets originally designed and built for either the Air Force or NASA and now being marketed commercially, such as the Titan III and Atlas I (formerly the Atlas-Centaur). Some vehicle types, like the Delta II and Atlas II, are simultaneously being produced under contracts to the Air Force and offered for commercial launch service.

An already strong legacy will gain greater luster in the next decade when a number of exciting and vital space science missions are launched for NASA on ELVs as part of the first phase. A Delta II booster procured through the Air Force will carry into space the Roentgen Satellite (ROSAT), an X-ray telescope that, once in orbit, will allow scientists to study such phenomena as the high X-ray luminosity of stars that otherwise appear to be identical to the sun. ROSAT will build on data provided by the High Energy Astronomy Observatory series, launched in the late 1970s on Atlas-Centaurs. West Germany is providing the telescope and the spacecraft, and the United Kingdom one of the focal instruments. The United States is procuring the launch vehicle and services, and also will provide one focal instrument.

NASA-sponsored exploration of the red planet, Mars, will resume with the Mars Observer mission in the early 1990s aboard a commercial Titan III rocket. Mars Observer will circle the planet for two years in a low, near-circular polar orbit, mapping the planetary surface as it changes with the seasons. Eight instruments will measure and investigate characteristics such as elements, minerals, cloud composition, and the nature of the Martian magnetic field.

Long-range plans call for NASA to purchase through the Air Force the most powerful American-made unmanned booster, the Titan IV-Centaur. The agency wants to launch two Titan IV-Centaurs in the mid-1990s as part of an ongoing program of solar system exploration.

The first planned mission is the Comet Rendezvous and Asteroid Flyby (CRAF), which will yield new insights into two types of smaller bodies in our solar system. En route to its rendezvous with the comet Knopff, CRAF will fly by the asteroid 449 Hamburga. It will take photographs and scientific measurements of the asteroid, which is only 55 miles (88.5 kilometers) in diameter. Once at Knopff, CRAF will spend three years flying alongside the comet. Scientists will be able to study a body of what could be some of the original matter left behind when our solar system was formed nearly 5 billion years ago.

The second Titan IV-Centaur is slated for the Cassini mission, named after a French-Italian astronomer who discovered several of Saturn's moons. Cassini will carry out a four-year tour of Saturn and its moons. It will also send a probe through the dense atmosphere surrounding Titan, the largest of Saturn's satellites, to collect data and provide a preliminary map of its surface. Cassini will be a joint effort between NASA and the European Space Agency.

In the second phase of the mixed fleet plan, ELV launch manufacturers will compete to launch a class of payload in a particular weight category—small, medium, intermediate and large. As in the transitional phase, the winning company will provide complete launch service, from building the ELV to launching it.

General Dynamics was the first ELV builder to receive an order under the second phase. In October 1987, NASA announced that it had chosen the Atlas I over Martin Marietta's Titan booster to launch a series of meteorological satellites for the National Oceanic and Atmospheric Administration (NOAA). At least three of the Geostationary Operational Environmental Satellite (GOES) spacecraft will be launched on Atlas Is from Launch Complex 36, and the number could reach five. Also in 1987, General Dynamics completed negotiations with NASA for use of the Launch Complex 36 facilities for commercial launches.

The two NASA centers which oversaw design and development of the Delta and Atlas rockets will still be involved with ELVs, but with a different slant. Goddard Space Flight Center is managing both competitive and non-competitive procurement of ELV launch services in the small payload class, which includes the Scout (not launched from Cape Canaveral), and medium weight payload class, which includes the Delta. In July 1989, Goddard announced that Delta manufacturer McDonnell Douglas had been competitively selected to negotiate a contract for three firm and 12 optional missions in the latter category.

Lewis Research Center will manage competitive and non-competitive procurement of ELVs capable of launching intermediate class payloads, which includes the Atlas-Centaur, and large class, which includes the Titan IV.

At KSC and Cape Canaveral, control of civilian unmanned launches is gradually being shifted from NASA to the private sector. In October 1988, Martin Marietta and NASA announced an agreement under which Martin will use some KSC facilities to support its Titan III commercial launches. Earlier the same year, a 28-year era drew to a close when NASA launched a Delta rocket for the last time from Launch Complex 17 on the Cape. In November the following year, the last Delta in the NASA inventory was launched on the West Coast. It carried the Cosmic Background Explorer (COBE) into orbit from Vandenberg AFB. In July 1988, Launch Complex 17 was formally turned over to the Air Force, which will allow Delta manufacturer McDonnell Douglas to conduct commercial launches from the same complex.

The Air Force also will assume control of Launch Complex 36; one of the two pads will be used by General Dynamics for commercial launches. In September 1989, NASA conducted its final launch of an Atlas-Centaur from the Cape. Atlas-Centaur 68, carrying the last in a series of Fleet Satellite Communications (FltSatCom) spacecraft, was originally scheduled for launch in 1987. The mission was postponed when the Centaur's liquid hydrogen tank was accidentally punctured by a work platform.

KSC will continue to be involved with ELV launches of U.S. civil payloads, but, as is the case with Goddard and Lewis, the slant will be different. In early 1989, KSC was assigned oversight responsibility for all unmanned launches carrying NASA payloads from both the Cape and Vandenberg AFB.

The NASA unmanned launch operations team has completed many historic launches. These space flights far exceed in number the manned missions of the Mercury, Gemini, Apollo, Skylab and Apollo-Soyuz programs. More than 300 unmanned launches were conducted from 1958 through the end of 1989, with a high rate of success. Unmanned spacecraft have returned volumes of scientific data, much of it not otherwise obtainable. The effects on scientific knowledge as a whole are incalculable. Technological benefits from applications spacecraft have provided better weather forecasting, accurate storm tracking, a superb international communications system that permits live television coverage from almost anywhere in the world, highly accurate navigation for ships and planes, and inventories of the resources of land and ocean. The unmanned space program already has more than repaid the nation's investment in time, money and technical talent as it enters its fourth decade.

The following table provides launch dates and brief descriptions of some of the more significant missions.

Launch Spacecraft	Launch Vehicle	Date	Mission Description
Vanguard 3	Vanguard SLV-7	9/18/59	Measured solar X-rays, determined environmental conditions in space, and explored Earth's magnetic field.
TIROS 1	Thor-Able-5	4/1/60	Relayed thousands of cloud pictures, demonstrating the feasibility of satellite observations in weather forecasting.
OSO-1	Delta-8	3/7/62	This first Sun observatory provided data on 75 solar flares, including scientific data not available from ground stations.
Telstar 1	Delta-11	7/10/62	First commercial communications satellite launched for the American Telephone & Telegraph Co.
Mariner 2	Atlas-Agena-6	8/27/62	First U.S. interplanetary probe to reach the planet Venus.
Syncom 2	Delta-20	7/26/63	First communications satellite in geosynchronous orbit.

Launch Spacecraft	Launch Vehicle	Date	Mission Description
Ranger 7	Atlas-Agena-9	7/28/64	First U.S. spacecraft to impact on the Moon; returned a series of photos and other data.
Pioneer 6	Delta-35	12/16/65	Launched into solar orbit; performed an early scientific study of the Sun from interplanetary space.
Surveyor 1	Atlas-Centaur-10	5/30/66	Performed first U.S. soft landing on the Moon, sending back thousands of excellent surface photographs.
OAO 2	Atlas-Centaur-20	12/7/68	Observed the Sun, stars, planets and nebulae in the ultraviolet spectrum, which does not reach the Earth's surface.
Mariner Mars 6	Atlas-Centaur-19	2/24/69	Passed within 2,000 miles (3,220 kilometers) of Mars' equatorial region, returning the first good photographs for the United States.
Pioneer 10	Atlas-Centaur-27	3/2/72	Performed flyby of Jupiter, returning first close-up photographs and measuring radiation emissions.
LANDSAT 1	Delta-89	7/23/72	First satellite to perform major assessment of Earth resources from outer space.
Copernicus	Atlas-Centaur-22	8/21/72	This OAO provided important information on the old questions about the origin and evolution of the universe.
Pioneer 11	Atlas-Centaur-30	4/5/73	Performed flyby of Jupiter and first flyby of Saturn.
Mariner Venus /Mercury	Atlas-Centaur-34	11/3/73	Performed a flyby of Venus, then continued on to Mercury and completed the first three flybys of that planet.

Launch Spacecraft	Launch Vehicle	Date	Mission Description
Helios 1	Titan III-Centaur-2	12/10/74	Approached the Sun to within outer solar corona and took density, temperature, velocity, and magnetic field measurements.
Viking 1	Titan III-Centaur-4	8/20/75	Placed an orbiter in orbit around Mars and a lander on the surface; obtained voluminous data in a search for life.
Viking 2	Titan III-Centaur-3	9/9/75	Same as Viking I except the lander was farther to the Martian north.
GOES 1	Delta-116	10/16/75	First weather satellite to photograph complete disk of the Earth every 30 minutes from geosynchronous orbit.
Voyager 2	Titan III-Centaur-7	8/20/77	Performed flybys of Jupiter, Saturn, Uranus, Neptune, and their moons.
Voyager 1	Titan III-Centaur-6	9/5/77	Performed flybys of Jupiter and Saturn with more scientific equipment than the Pioneers; photographed moons and planets; discovered Jupiter's ring, and volcanoes on Jupiter's moon Io.
Pioneer Venus Orbiter	Atlas-Centaur-50	5/20/78	Placed in orbit around Venus to study the atmosphere and surface; compiled radar maps of surface features.
Pioneer Venus Multiprobe	Atlas-Centaur-51	8/8/78	Sent four probes into the Venusian atmosphere five days after the orbiter arrived; returned much useful data.
ISEE 3	Delta-144	8/12/78	First satellite in a "halo" orbit, poised between Sun and Earth; studied solar wind and gamma ray bursts.

chapter 3

MERCURY and GEMINI

The ladder NASA climbed to reach the Moon had three rungs of achievement—the Mercury, Gemini and Apollo programs. The first program, Project Mercury, was initiated on Oct. 7, 1958, just six days after the founding of NASA. Its objective was to orbit and retrieve a manned Earth satellite.

In mid-September of 1958 the National Advisory Committee for Aeronautics (NACA), NASA's predecessor agency, and the Advanced Research Projects Agency had established a Joint Manned Satellite Panel. After NASA took the place of NACA, it organized a Space Task Group at its Langley Research Center to direct the program. The group was responsible directly to NASA Headquarters.

Although the X-1 and other research rocket-planes had taken American pilots to the fringes of space and although the principal problems posed by living in weightlessness (eating, drinking, and elimination) were easily solved, there remained physiological and psychological unknowns. Radiation, isolation, and re-entry stresses had to be overcome before people could venture into orbit. Moreover, rockets had to be made more reliable, or "man-rated."

The rocket chosen to carry the Mercury payload into orbit was an Atlas Intercontinental Ballistic Missile (ICBM). "Big Joe" was one such missile, capped with a full-scale Mercury spacecraft. Launched on Sept. 9, 1959, it tested the heat shield which protected the capsule from the searing temperatures endured during re-entry. The capsule survived, and an autopsy on the heat shield proved its structural integrity.

"Little Joe" was a special booster, simple and relatively inexpensive, which carried a primate passenger named Sam through 3 minutes of weightlessness back to safety on Earth. The rocketing rhesus monkey's flight also tested the worldwide tracking network for Mercury.

Other early developmental flights in the Mercury program were not so successful. Mercury-Atlas 1 exploded one minute into its flight—causing the program a six-month delay. Mercury-Redstone 1 had a very short liftoff. It rose 4 or 5 inches (10 to 13 centimeters) before settling back on its fins, while the escape tower launched—without its attached capsule. On a manned mission, the tower was supposed to carry an astronaut to safety if the flight were aborted.

The repeat flight of the Redstone (1A), 28 days later, was a success, as was the Mercury-Redstone 2 flight of the astrochimp Ham, launched in January 1961. On his 16-minute suborbital flight, Ham performed a series of tasks for which he had been trained, functioning in space like a normal chimpanzee.

The Redstone rocket, though not powerful enough to place the Mercury capsule in orbit, was selected for two suborbital manned flights. On May 5, 1961, astronaut Alan B. Shepard Jr. became America's first man in space. Inclement weather, a faulty inverter in the electrical system and a computer problem caused slight delays in the launch. But the mission of Mercury-Redstone 3 went smoothly after it finally left the pad, with Shepard demonstrating no ill effects from either weightlessness or gravitational stresses.

Shepard's 15-minute suborbital flight in Freedom 7 took him 302 miles (486 kilometers) from Cape Canaveral. He was weightless for 5 minutes, cramped with equipment in a space capsule 9.5 feet (2.9 meters) high and 6 feet (1.8 meters) in diameter at its base.

The next suborbital flight, flown by Virgil I. Grissom, also lasted 15 minutes. The launch on July 21 had one major hold when a gantry technician discovered that one of the 70 bolts of the space capsule escape latch was improperly aligned. The problem was corrected and the countdown continued.

A serious emergency occurred during splashdown, when the capsule hatch opened and seawater flooded in. Grissom was forced to abandon ship. His Liberty Bell 7 sank to the bottom of the ocean.

Redstone, one of the most reliable of the U.S. liquid fuel rockets, was chosen for the first suborbital flights in the pioneering Mercury program.

When the Redstone engine cut off during this unmanned Mercury launch attempt, the escape tower staged its own unscheduled liftoff, leaving the spacecraft parked on the pad.

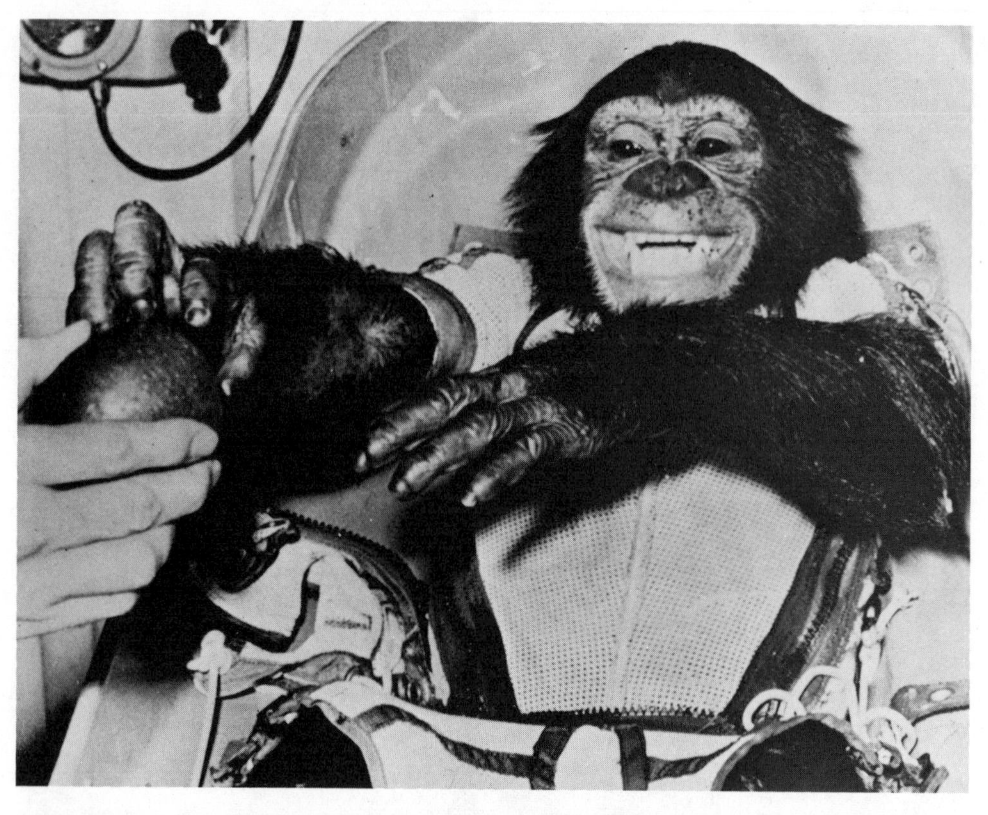

Chimpanzees such as "Ham," shown receiving an apple at the conclusion of a 420-mile (676-kilometer) Mercury flight, helped to accumulate needed data on effects of space flight.

Mercury-Atlas 3 was supposed to carry a "mechanical astronaut" into orbit, but its booster failed. In September, Mercury-Atlas 4 was launched on a successful one-orbit mission. Its instruments monitored levels of noise, vibration and radiation. Mercury-Atlas 5 was a two-orbit chimpanzee flight. The chimp Enos, Hebrew or Greek for "man," overheated but was recovered unharmed.

John H. Glenn Jr. was selected as the first U.S. astronaut who would attempt to enter orbit, propelled by an Atlas vehicle. The launching was postponed repeatedly because of technical problems in the fuel tanks, bad weather, a slipped thermistor on Glenn's suit, a broken hatch bolt, a stuck valve, too little propellant in the booster's tank, and a power failure at the Bermuda tracking station.

These problems were all solved and on Feb. 20, 1962, almost a month past the initial launch date, Friendship 7, with Glenn aboard, blasted into orbit. On the second and third orbits, because of difficulty with the automatic stabilization and control system, Glenn had to take manual control—to fly-by-wire in space jargon—and so had to omit many of the assigned observations.

Alan Shepard became the first American in space on May 5, 1961, when this Mercury-Redstone rocket carried him on a 15-minute suborbital flight.

John Glenn became the first American to fly an orbital mission in space in February 1962.

A serious problem occurred during re-entry. Data telemetered to Earth indicated that the heat shield was no longer locked in place. A spent retropackage was left underneath the shield in hope that it would keep it in place during the blazing re-entry from orbit. Glenn survived, and the heat shield was found intact; it was the telemetered data that had been wrong.

M. Scott Carpenter flew three orbits in Aurora 7 in May 1962, with more pilot control of the mission, including inverted flight (pilot's head oriented toward Earth). Walter M. Schirra Jr. completed six orbits in Sigma 7 in October 1962. And in May 1963, L. Gordon Cooper made the last flight of the Mercury program, called the "daylong" mission, in Faith 7. He orbited 22 times and splashed down 34 hours and 20 minutes after liftoff.

One of the lessons learned in Mercury was how much time final launch preparations took at the spaceport. NASA decided to build an automated digital system to reduce the spacecraft's time on the flightline in the future. This was called the Acceptance Checkout Equipment (ACE).

After nearly a month of delays due to technical problems, an Atlas vehicle lifted Glenn's Friendship 7 into orbit for a historic flight and a safe return.

Built on the foundations Mercury had already established, the Gemini program was the next major step to the Moon. Gemini, as the name reflects, was a two-man spacecraft, far more sophisticated than the Mercury capsule, although they looked much alike. For a while it was known as the Mercury Mark II. Gemini was launched by a Titan II missile, developed by the Air Force.

Gemini's accomplishments included the first rendezvous of one spacecraft with another—an extremely difficult maneuver because of the astronavigation involved; the first docking of a spacecraft to another propulsive stage, and use of that stage to propel the spacecraft into a higher orbit; and the first human travel into the Earth's radiation belts. With this program, the United States surpassed the Soviet Union in manned space flight.

In 1961, President Kennedy had committed the nation to putting an astronaut on the Moon before the end of the decade. Gemini would demonstrate capabilities needed for lunar flight: rendezvous and docking, duration of mission for up to two weeks, and controlled landing.

Gemini-Titan 1, an unmanned launch vehicle, gave an almost flawless performance in March 1964. So did the second unmanned Gemini-Titan in January of the next year, after a myriad of problems prior to launch. In August 1964, lightning struck Complex 19 at the Cape. Then Hurricane Cleo brushed the coast, followed by scares from her sisters Dora and Ethel. The countdown finally started on Dec. 8. The first-stage engines ignited and sprang into life, only to be shut down one second later because the vehicle lost hydraulic power in its primary control unit. For the first time, launch crews had to drain hypergolics—propellants that ignite upon contact with each other—from a vehicle at the pad. Launch on Jan. 19, 1965, was successful.

All consoles manned, Mercury Control is in full readiness for a Mercury-Redstone 4 flight, one of the pathfinding missions that opened the way for the Gemini program.

On March 23, after a 22-month gap in U.S. manned missions, Grissom and John Young were launched on a three-orbit flight on Gemini-Titan 3. Grissom, whose first spacecraft had sunk to the ocean bottom, nicknamed this one "Molly Brown" after the "unsinkable" heroine of a Broadway musical. The astronauts did the first maneuvering in orbit with Molly Brown.

On the Gemini 4 mission flown by James A. McDivitt and Edward H. White II, White became the first American to "walk" in space when he performed a 20-minute extravehicular activity. White wore a space suit which weighed 31 pounds (14 kilograms) and contained 18 layers of material to protect him from heat, cold, and meteoroids, and to maintain internal space suit pressure.

The Gemini 5 mission in August 1965, with Cooper and Charles Conrad Jr. at the controls, was an eight-day voyage. It was the first flight to use fuel cells instead of storage batteries to produce electricity. The cells provided electrical power through the reaction of liquid hydrogen and liquid oxygen, a method which allowed manned flights to continue longer.

Gemini 6 was to be a rendezvous mission, incorporating an Atlas-Agena as a target vehicle. But the Agena exploded soon after launch in October 1965.

Two McDonnell officials, Walter Burke, the spacecraft chief, and John Yardley, his deputy, suggested performing a rendezvous mission by flying Gemini 6 and 7, two manned vehicles, at the same time. On Dec. 4, 1965, Gemini 7 was launched, carrying astronauts Frank Borman and James A. Lovell Jr. on their 14-day mission. On Dec. 15, Gemini 6-A, with Schirra and Thomas P. Stafford, followed the path to orbit. That same day it rendezvoused with Gemini 7. This was the first time the tracking network had simultaneously tracked and acquired information from two orbiting manned spacecraft.

During the Gemini 4 mission, Edward H. White II became the first American to "walk" in space, spending 20 minutes outside the capsule in a space suit weighing 31 pounds (14 kilograms).

The Atlas-Agena for Gemini 8 was launched into a circular orbit on March 16, 1966. The Gemini, with astronauts Neil A. Armstrong and David R. Scott, followed, and rendezvoused and docked with the Agena. Soon afterward, the docked vehicles began rolling in a mad whirl, forcing Armstrong to undock. The Gemini continued revolving; one of the maneuver thrusters had stuck open. Armstrong turned off the maneuvering system and turned on the re-entry control system, which restored pilot control. Cutting the mission short, they landed the same day.

Tragedy struck the space program when astronauts Elliot M. See and Charles A. Bassett II, scheduled to fly Gemini 9, died in a crash of a T-38 jet aircraft. Flying into St. Louis in poor weather conditions, they crashed into the roof of a building near the airfield. Ironically, this facility, McDonnell Building 101, housed the spacecraft they were to ride into orbit. Gemini 9 backups Stafford and Eugene A. Cernan replaced See and Bassett on Gemini 9. Their target vehicle, an Atlas-Agena, lifted off on May 17, 1966—and immediately flipped over into a nose dive.

An alternate target was available. After the Agena explosion of October 1965, NASA had ordered a backup Atlas called the Augmented Target Docking Adapter (ATDA). It consisted of a target docking adapter bolted to the rendezvous and recovery section of a Gemini and fitted to the Atlas. On June 1, 1966, the ATDA reached orbit, but a launch shroud protecting the docking port had failed to jettison. Gemini 9-A lifted off on June 3, and rendezvoused with the ATDA. With the shroud half-opened, it looked to Stafford like an "angry alligator." Docking was impossible. Cernan had an extravehicular

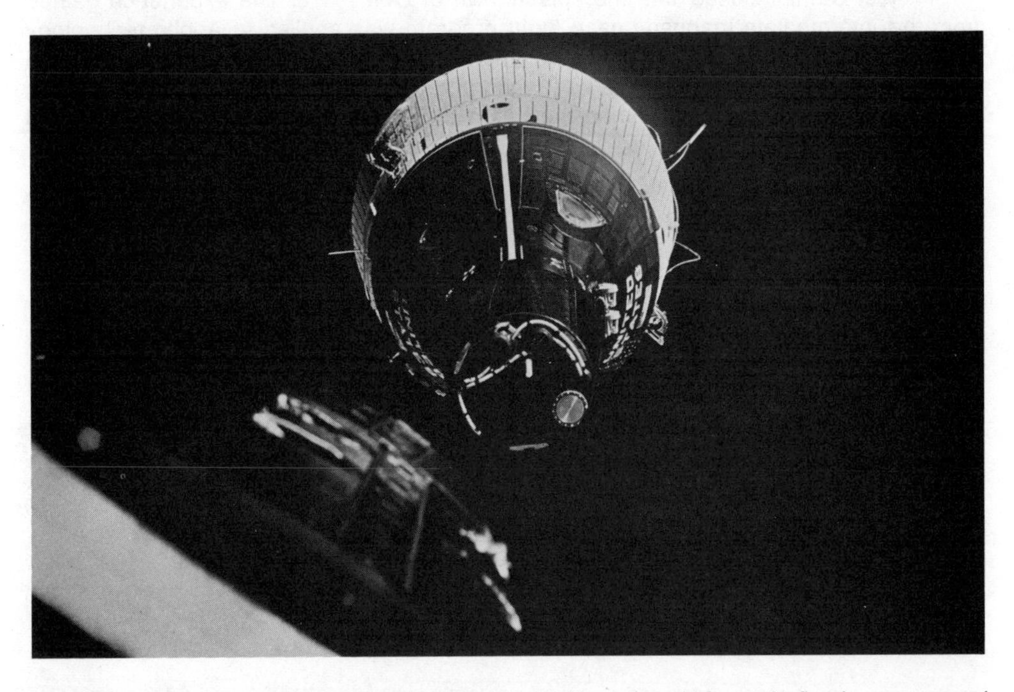

The orbital rendezvous of Gemini 7 and Gemini 6A, pictured from aboard 6A, was the first time two manned spacecraft had been tracked simultaneously.

activity scheduled and could have tried to remove the shroud, but the possibility of tearing his space suit on the jagged edges made the endeavor too risky.

Young and Michael Collins had better luck with their mission. Gemini 10 and its corresponding Atlas-Agena rose into orbit on July 18, 1966. Young docked with the Agena and fired its main engine to propel both vehicles into a higher orbit. Then the Agena propelled the spacecraft to rendezvous with Gemini 8's Agena. After releasing their own Agena from bondage, they approached the other within 6.5 feet (two meters). Collins left the spacecraft, drifted over to the Agena and removed one of its experiments.

A mere two seconds was the length of the launch window for Gemini 11—the shortest in the Gemini program. On Sept. 12, Conrad and Richard F. Gordon Jr., in their Gemini spacecraft, were launched for a first-orbit rendezvous with an Agena stage, launched the same day as the usual Atlas booster. After rendezvous, they engaged in docking and undocking practice. Astronaut Gordon took a "space walk" with a tether attachment between the undocked Gemini and Agena. Leaving space, their automatic re-entry gave them a very accurate landing.

Nov. 12, 1966, saw the beginning of the end of the Gemini program. As Lovell and Edwin E. Aldrin Jr. took the pad elevator up to Gemini 12, they carried signs on their backs, "The" and "End." After launch of the manned spacecraft and its Agena target, the Gemini's radar failed. They were able to rendezvous anyway, using sextant measurements, thanks to the expertise of Aldrin, whose nickname was "Dr. Rendezvous."

Project Gemini closed with the splashdown of Gemini 12. The experience gained from the intermediate manned space flight program was applied to Apollo technology even before the Gemini program ended. As modifications were made to upcoming Gemini hardware, those same innovations were incorporated into the design of Apollo. Apollo engineers and managers welcomed the baton they were handed and began the final stretch to the Moon.

chapter 4

GENESIS of APOLLO

When President John F. Kennedy announced the national objective of landing a man on the Moon within the decade of the 1960s, NASA faced the unprecedented task of transforming this goal into reality. At the outset, the means were unclear. The spacecraft that would carry American astronauts to the Moon and back existed only as a theoretical concept, tentatively called Apollo. The powerful rocket that would be needed to launch the spacecraft with sufficient velocity to escape Earth's gravity was only a few lines on an engineer's scratch pad. The vast support, checkout and launch facilities required to launch the daring lunar voyagers had to be designed, built and activated.

Intense effort by NASA's rapidly growing government-industry-university team gradually filled in the grand design. The ambitious program called for launching three astronauts into lunar orbit. Two of them would then pilot a smaller spacecraft to land on the lunar surface. The third would orbit the Moon in the mother ship while his companions completed the initial lunar exploration. Then they would rendezvous with his craft and the trio would return to Earth in the mother ship. This basic decision fixed size and weight parameters for the booster rocket, which in turn dictated the dimensions and scope of the facilities needed to service and launch the rocket and its two spacecraft.

The Marshall Space Flight Center was given the responsibility for providing the launch vehicle for the program. Called the Saturn V, this vehicle consisted of three separate stages and an instrument unit, atop which sat the spacecraft.

The Boeing Co. won the contract to fabricate the large first stage. It contained five engines, each generating 1.5 million pounds (6.7 million newtons) of thrust. The Rocketdyne Division of North American Aviation furnished these engines. North American also received the entire second stage contract, calling for a rocket stage equipped with five engines, each with 200,000 pounds (900,000 newtons) thrust capacity. Douglas Aircraft Co. won the competition for the third stage, equipped with a single 200,000-pound (900,000-newton) thrust engine with a restart capability, required for the lunar mission.

IBM was selected to provide the instrument unit, containing a computer and the electronic control systems which steered the vehicle and corrected its trajectory in flight.

While Marshall assembled its industrial partners, the Manned Spacecraft Center (now the Lyndon B. Johnson Space Center) conducted the competition for spacecraft contractors. North American won the award to develop the Apollo command and service modules which would carry the astronauts from Earth to lunar orbit and back. Grumman Aircraft Engineering Corp. received the contract to fabricate the lunar landing spacecraft.

The Massachusetts Institute of Technology provided the spacecraft guidance system. General Electric built the automated Acceptance Checkout Equipment to test the command, service, and lunar landing modules. In all, some 300,000 men and women, in hundreds of firms across the nation, shared in the total program.

Among the responsibilities of the Kennedy Space Center was the requirement to provide the launch site for the Saturn V. KSC's director, Dr. Debus, provided engineers with pencil sketches depicting innovative concepts. From these rough drawings emerged the structures which comprised the launch facility. The ground servicing facilities had to match perfectly with vehicles that, at this time, existed only on paper.

Space Center planners were guided by four fundamental considerations:

- Only those activities essential to checkout, mating, testing, erection, and launch of the Apollo-Saturn V would be conducted at the launch complex.

- All supporting activities, and those concerned with testing Apollo modules before mating, would be performed in an area separate from the launch complex.

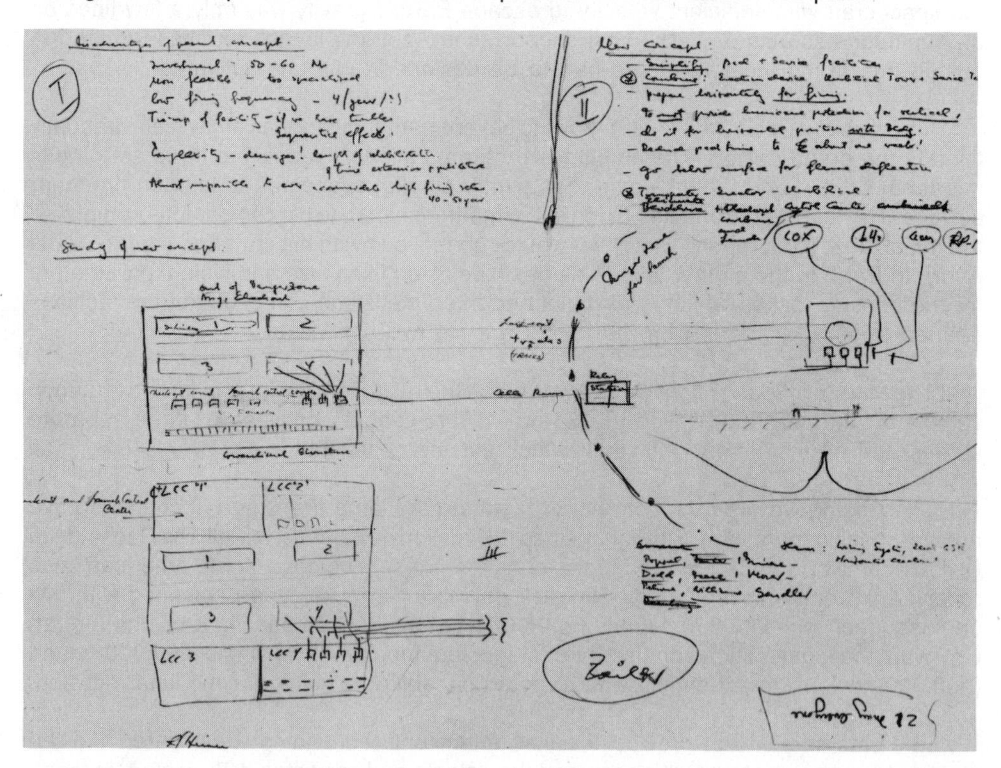

Dr. Kurt Debus' original sketch of an assembly building and launch control center compared the advantages of assembling a space vehicle on the launch pad versus assembling it in a protective structure and moving it to the pad for launch.

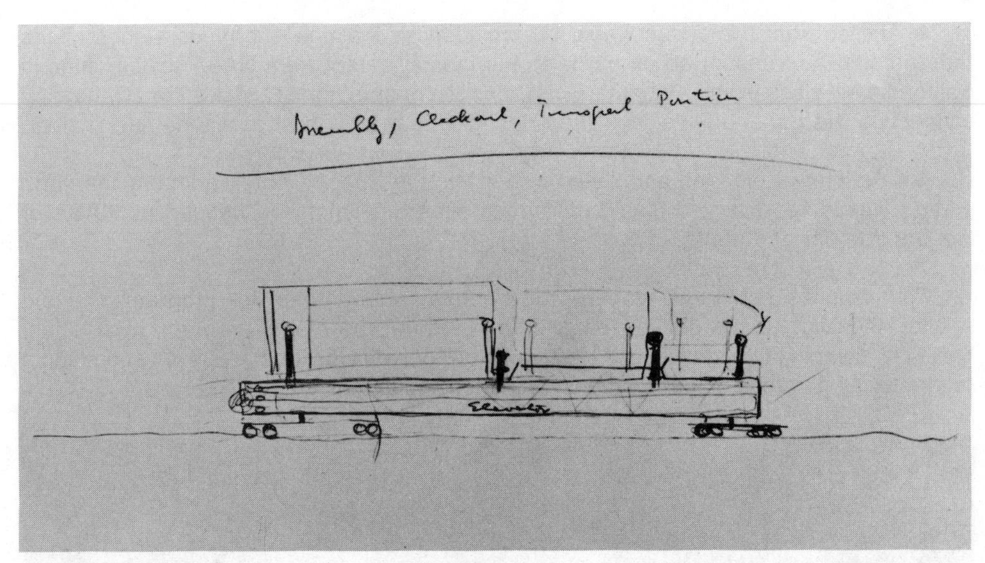

Dr. Debus' early concept of a transporter to carry assembled space vehicles to the pad, pictured above, was later changed from a horizontal to a vertical mode.

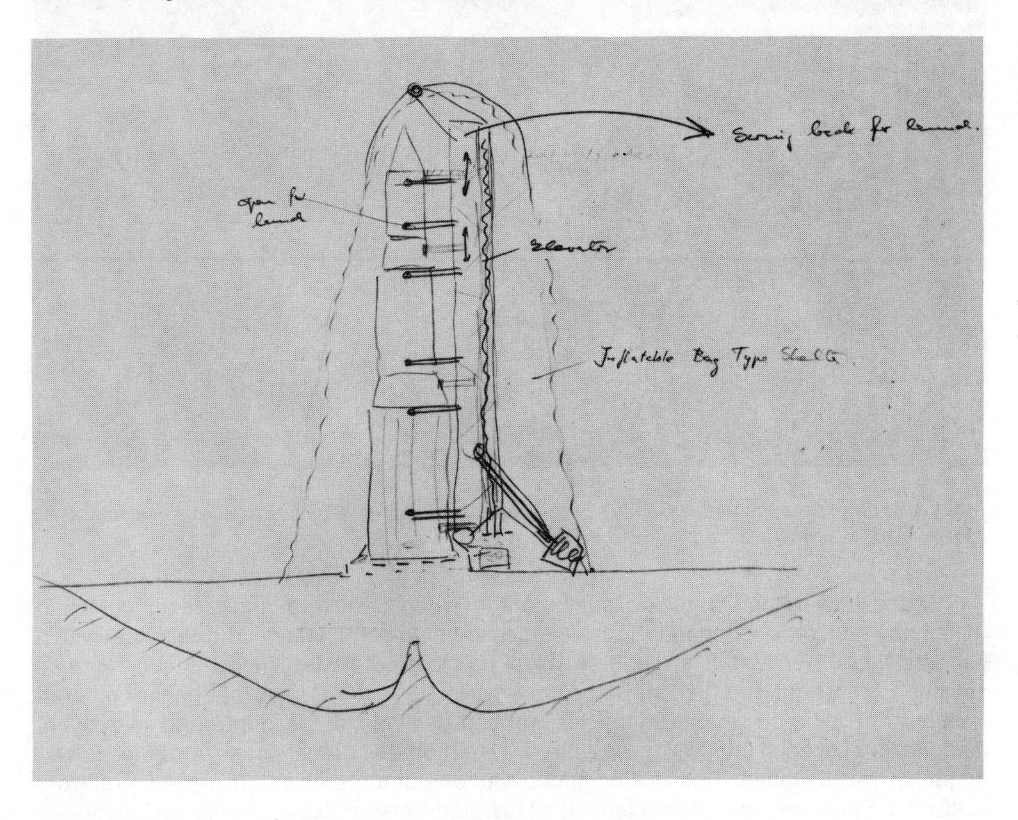

This is how Dr. Debus visualized the Apollo-Saturn V at the launch site.

● Access from the ocean would be provided so that the enormous stages of the Saturn could be transported by barge to the launch center from NASA testing sites in California, Mississippi and Alabama. This required construction of a lock at Port Canaveral connecting the Banana River and the Atlantic, and dredging of a channel in the river.

● One-third of the land and water area within the Space Center, primarily the north portion, would be reserved for future launch sites that might be needed by NASA or the Department of Defense.

The facilities had to be capable of meeting requirements for programs beyond Apollo—then not clearly defined—as well as those of the Apollo program. In addition, the size, complexity, and cost of Apollo-Saturn V vehicles and those of succeeding programs demanded radical changes in launch techniques.

A new kind of vessel appeared in Florida's coastal waters as barges carrying stages of the giant Saturn booster made their way to Complex 39 at KSC.

Dating back to the first rocket launched from Cape Canaveral, those techniques had remained relatively unchanged. Rocket stages underwent inspection and checkout within hangars while horizontal. Then they were transported to the launch pads for final assembly, checkout and fueling. Since the environment of the checkout hangar differed from that of the pad, many of the test procedures had to be duplicated to prevent malfunctions. At each pad, which was a heavily reinforced concrete base, a service tower, or gantry, was installed. This was equipped with one or more elevators and several work platforms. Engineers and technicians used the platforms for access to the vehicle during the prelaunch preparations. An umbilical mast, or tower, provided support for the lines that carried fuel and power into the rocket from ground sources.

New facilities, of a kind not existing anywhere else, were needed--and Launch Complex 39 was conceived out of engineering initiative and construction know-how.

Close to the pad was a concrete-and-steel shelter, or blockhouse, housing the control systems. It also protected the launch crews during hazardous tests and launches. Cape Canaveral blockhouses were only a few hundred feet from the launch pads. At the time of construction, greater separation was not feasible due to the state of the art in electronics technology: signal degradation between the rocket and control systems in the blockhouse left no alternative but to keep them in close proximity.

The assembly-at-the-pad concept of launch preparations tied up the total facility for long periods of time. Severe weather could halt work or even require disassembling and moving the vehicle back to the hangar for safekeeping.

Experience convinced Debus that a more efficient method had to be devised to cope with vehicles like the Apollo-Saturn V. His early sketches expressed a mobile concept. That is, the rocket would be assembled and checked out with the spacecraft in the protective environment of a building, then transferred to the launch site when almost ready for flight. This would prevent interruptions during checkout, mating and erection; provide greater assurance of test integrity; and materially increase the launch rate from the same pad. In an emergency such as a hurricane alert, the means of transporting the vehicle to the pad could also carry the vehicle back to the hangar, preserving all the vital connections, and return it to the firing site when the storm had passed.

This was the challenge for the engineers, to create facilities unlike any elsewhere in the world, for a program that had no historical precedent. The result was Launch Complex 39.

chapter 5

CONSTRUCTING the SPACEPORT

Launch Complex 39, originally designed to accommodate Apollo-Saturn V space vehicles, taxed the ingenuity of its construction team. Architectural Forum magazine described the four-year, $800 million-project as "one of the most awesome construction jobs ever attempted by Earth-bound men."

Its principal features included:

● A hangar big enough to enclose four Apollo-Saturn V space vehicles, each standing 363 feet (111 meters) tall and measuring up to 33 feet (10 meters) in diameter.

● Movable platforms on which the rockets were assembled and transported to the launch pad, and from which they were launched.

● A system of carrying rockets and launchers, weighing 12 million pounds (5.44 million kilograms) to and from the firing site.

● A movable service structure 45 stories tall that permitted technicians to complete preparations at the launch pad.

● A control center from which preparations were monitored and controlled through launch.

KSC planners drew up the requirements while the Army Corps of Engineers created a new management office, the Canaveral District, to supervise brick and mortar construction contracts for NASA. The Florida East Coast Railroad built a causeway across the Indian River, connecting the spaceport and the mainland in order to haul freight directly to the building site. Structural steel, rolled in northern mills, came by truck from Tampa, on Florida's west coast, where it had received special treatment against corrosion.

Late in 1962, sufficient planning data had been collected from rocket and spacecraft designers to begin detailed drawings for the huge hangar, to be designated the Vehicle Assembly Building (VAB). The Apollo-Saturn V was by far the largest space vehicle in the world, and the structure originally built to accommodate it is one of the world's largest in volume. The building encloses 130 million cubic feet (3.68 million cubic meters) and covers eight acres (3.2 hectares).

Debus conceived a cruciform shape in his early sketches of the hangar. Another plan proposed lining up checkout bays in a single row that would have resulted in a narrow, slablike shape. NASA chose a back-to-back layout, with a transfer aisle between rows of checkout and assembly bays. The boxlike, rectangular shape that resulted was both more economical and stronger, capable of withstanding hurricane-force winds.

More than 98,000 tons (89,000 metric tons) of steel were used to construct the VAB. In all, 45,000 separate pieces of steel weighing from 150 to 72,000 pounds (68 to 32,659 kilograms) were required. More than 1 million bolts secured the steel members. The paperwork for the VAB project alone was monumental. Approximately 2,500 separate drawings were submitted. It took an entire railroad freight car to move the final specifications printed for the bidders.

Incorporating 98,590 tons (89,717 metric tons) of steel, and resting on 128 miles (206 kilometers) of 16-inch (40.6-centimeter) diameter steel pipe, the Vehicle Assembly Building takes form at Complex 39.

Clearing and filling operations for the VAB began in November 1962. Since the average height of the terrain was barely above sea level, dredges were brought to the nearby Banana River to pump fill onto the land. This dredging raised the land level to almost seven feet (2.1 meters) by July 1963 and, at the same time, opened up the channel and turning basin for the barges that would transport the rocket stages. In all, 1,500,000 cubic yards (1,140,000 cubic meters) of soil were pumped from the river bottom to the site.

The foundation presented unusual problems. Tests showed a three-foot (one-meter) limestone shelf 118 feet (36 meters) below the surface. The layers below that were silt overlaying bedrock at a depth of 160 feet (49 meters). Core borings revealed petrified wood, carbon-dated to be 25,000 years old. It was decided that the massive VAB would rest upon steel pipe pilings, 16 inches (41 centimeters) in diameter and 3/8 inch (10 millimeters) thick, driven into the limestone bedrock. The foundation contractor spent six months driving 4,225 pilings through the upper limestone layer. When the job was complete, 128 miles (206 kilometers) of steel pipe had been buried. Each pile was filled with sand to within one foot (30.5 centimeters) of the top, and concrete caps were poured around clusters of from six to 20 pilings. Since the piling penetrated a salty chemical solution below the surface, there was a tendency for electrolysis to occur. Cathodic protection had to be applied to neutralize the current; otherwise, the foundation would have corroded and the VAB could claim to be the world's largest wet cell battery.

To anchor the building securely, 30,000 cubic yards (23,000 cubic meters) of concrete formed the floor slab.

The building was structurally completed in June 1965 when a large steel beam painted white for the purpose was autographed by construction workers, NASA, and Corps of Engineers employees, and hoisted into place at the traditional topping-out ceremony.

Looming above the flat terrain, the VAB does not forcibly impress visitors because there is no other nearby tall structure to provide a comparison. Not until one enters the transfer aisle and looks up to the supporting beams under the roof can the dimensions of the building be grasped. It measures 716 feet (218 meters) in length, twice the size of a football field, and 518 feet (158 meters) in width. It is 525 feet (160 meters) high. Under strong wind pressure the structure can sway 12 inches (30.5 centimeters).

The VAB is divided into two main sections. The high bay portion contains four bays, each large enough to accommodate a mobile launcher carrying a fully assembled space vehicle.

For Apollo, the low bay portion contained eight cells used for preparation and checkout of the second and third stages of the Saturn V. Each cell was a structural steel assembly equipped with work platforms that opened to receive the stage and then enclosed it. Mechanical and electrical systems in each cell permitted simulation of stage interfaces and operations with other stages, as well as with the instrument unit.

Three high bays were fully equipped for the Apollo-Saturn V vehicles. The fourth was reserved for other vehicles that might later be required. Any vehicle that can fit within the high bay doors can be prepared for launch in the building. Five pairs of extensible work platforms were installed in each operational bay. Some were one story tall, others two stories, and some as high as a three-story building. The platforms were installed on the sides of the bay and could be adjusted upward or downward, in and out, encircling the Saturn V stages during checkout and preparation. They were retracted against the walls when the vehicle was moved to the firing site.

There are 141 lifting devices in the building ranging from one-ton (0.9-metric ton) hoists to two 250-ton (227-metric ton) bridge cranes with hook heights of 462 feet (141 meters). The largest cranes were used to lift and mate rocket stages, and were required to hold the load up to 50 minutes without moving more than 1/32 inch (eight millimeters). Electronic controls enabled operators to move the cranes in all directions at speeds as low as 1/10 foot (2.5 millimeters) per minute--the remarkably slow speed was required in mating Saturn V stages.

Precision control to the nth degree is required as this 250-ton (228-metric ton) capacity crane gently lowers the Saturn V first stage to the deck of the mobile launcher.

Gigantic doors in the shape of an inverted "T" form the outer wall of the high bays. Four leaves slide horizontally to close the lower portion. Seven leaves above them operate vertically. The doors are taller than St. Peter's Basilica, the Statue of Liberty, or Chicago's Wrigley Tower. It takes 45 minutes to open a door.

To prevent condensation and fogging within the structure, a gravity ventilation system forces a complete change of air every hour through 125 ventilators placed on the roof. Selected work areas within the building are air-conditioned.

Much of the office space in the upper levels housed the checkout instrumentation of the stages. Sixteen high-speed elevators served the 3,000 employees working in the VAB during the Apollo program. Each contractor occupied floors convenient to his stage—Boeing next to the first stage, North American the next levels, Douglas Aircraft above them, and IBM near the instrumentation unit.

Operations during the Apollo program generally followed this sequence:

- A mobile launcher was carried into a high bay and parked on support pedestals by the crawler-transporter.

- The first stage of the Saturn V rocket was off-loaded from a barge at the turning basin and hauled, horizontally, into the low bay on a transporter.

- Moved into the transfer aisle, the stage was positioned in front of a high bay, picked up by the 250-ton (227-metric ton) crane, tilted into a vertical position, raised 195 feet (59 meters) and then passed through an opening in the bay trusswork and gently lowered to the deck of the launcher.

- Second and third stages were brought into the low bay and placed in checkout cells, those on the east side holding the larger second stage, while the smaller stages were placed in the west cells.

- Inspection, checkout, and interface checks of all three stages proceeded concurrently. Then the second stage was removed by crane, carried into the transfer aisle, raised vertically through the bay opening, and placed atop the first stage. The same procedure was followed to stack the third stage, and finally the instrument unit topped off the launch vehicle.

- Each stage was electronically connected with a computer in the base of the mobile launcher, which communicated with a computer in the Launch Control Center firing room. Checkout instrumentation located in adjoining laboratories near the bay was hooked up, and the preparation work continued.

- Meanwhile, the spacecraft modules had been put through similar testing in the Industrial Area. When these tests had been satisfactorily completed, the spacecraft was assembled, carried to the VAB on a low-bed transporter, and moved into the transfer aisle. The powerful crane lifted the spacecraft through the opening in the bay and placed it atop the Saturn V.

- When the preparations in the VAB had been completed, the crawler-transporter returned, the bay doors opened, and the machine picked up the mobile launcher and rocket for the transfer to the launch pad.

Connected with the VAB on the southeast by a cableway and personnel access corridor three floors above the ground is the Launch Control Center. The Launch Control Center, which differs completely in shape and construction from the squat, conical blockhouses that dot Cape Canaveral, contains four levels. Offices, service areas, originally a cafeteria and a control center for launch support functions occupy the first level. The second level houses telemetry and recording equipment, instrumentation and data reduction facilities. The third level divides into four firing rooms or control rooms.

During Apollo, three of the four firing rooms contained a computer room, mission control room, test conductors platform, visitors gallery and adjacent offices. (The fourth firing room was never equipped with checkout consoles.) The fourth level contains offices and mechanical equipment.

From the firing rooms in the Launch Control Center, the launch crews monitored and controlled the multiple technical operations performed in the course of the checkout, mating, testing, fueling and launching of the powerful Apollo systems.

An Apollo spacecraft is checked out from this room at KSC, using the automated Acceptance Checkout Equipment developed for the program.

Two separate computer systems were employed. The Acceptance Checkout Equipment, designed and operated by General Electric Co., was used for Apollo spacecraft. Housed in the KSC Industrial Area five miles (eight kilometers) away, the equipment allowed engineers to monitor 24,000 samples of spacecraft test data per second. The Launch Control Center housed the Saturn ground computer system, which had two RCA 110A computers—one in the active firing room, the other in the base of the mobile launcher. The measuring program for the Saturn V rocket checked 2,728 discrete functions, providing verification that critical components were operating properly during prelaunch tests and launch.

The same contractor that built a stage also provided the engineers who manned the consoles controlling its preparation for flight. All functioned under the direction of KSC's government employees.

Launch crews in old Cape Canaveral blockhouses viewed launches on closed-circuit television monitors or through periscopes. The Apollo-Saturn V firing room crews watched rockets lift off through huge windows in the east wall. The three firing rooms equipped for Apollo were connected with the three bays of the VAB also equipped for these missions. Using the same instrumentation for prelaunch checkout in the VAB and for fueling and launching at the pad assured uniform standards of measurement, regardless of where the space vehicle happened to be at the time. The fourth Launch Control Center firing room functioned as a planning center for daily work schedule discussions, where progress of the checkout was plotted and the government/contractor engineers exchanged schedule information.

South of the Launch Control Center is the barge terminal, consisting of a turning basin, dock, barge slips and material handling area. Here the barges arrived with the first and second stages of the Saturn V, each too long to be shipped by air or land. The barges crossed the Gulf of Mexico, circled the southern tip of the Florida Keys and proceeded up the Atlantic coastline to Port Canaveral, through the Banana River lock and channel to the receiving dock. The Saturn V third stage, instrument unit, and Apollo spacecraft arrived by air in special cargo planes called "Guppies."

The mobile launchers were the key to the mobility characterizing Launch Complex 39 operations. The mobile launchers were transportable steel structures which moved erected Saturn V vehicles to the launch pad. Three identical launchers were built for the three equipped high bays of the VAB. Erection of structural steel for the first launcher began in July 1963, and the third was topped out in March 1965.

Capped by a 25-ton (23-metric ton) hammerhead crane, a mobile launcher stood 445 feet (136 meters) tall, and weighed a total of 6,300 tons (5,700 metric tons) when carrying an unfueled Apollo-Saturn V.

A mobile launcher had two functional areas, a launcher base and an umbilical tower. The launcher base was a two-story steel platform about one-half acre (0.2 hectare) in size on which a Saturn V was assembled, transported to the launch site, and launched. Within the base was a computer linked with the computer in the firing room of the Launch Control Center.

A red umbilical structure towered over the base. It carried nine swing arms, or bridges, for direct access to the space vehicle; 17 work platforms; and distribution lines for propellant, pneumatic, electrical, and instrumentation systems. Swing arms were mechanical bridges operated by hydraulic systems. They provided access to the Apollo-Saturn V during assembly, checkout and servicing. Arm number 9, 320 feet (98 meters) above the base, connected with the Apollo spacecraft, and was used by the astronauts to enter the command module.

Swing arm technology had been successfully used at Complexes 34 and 37 on Cape Canaveral for the smaller Saturn I vehicles. Those required at Complex 39 were much heavier and proposed difficult engineering problems. They supported propellant lines through which fuel entered tanks of all three Saturn V powered stages, as well as supporting electrical and pneumatic feeds from the ground. Four arms could be disconnected prior to launch, but five others carried lines that could not disengage until the rocket began to move. In the jargon of the Space Center, they had to swing clear at "first motion." They retracted in two to five seconds to provide room for the wider, lower stages of the giant Saturn V to pass as it rose. As a further precaution, the rocket steered on an angle away from the umbilicals.

Giant hold-down arms, whose name accurately describes their function, were mounted on the surface of the base to support and restrain the vehicle. They held the rocket during the first 8.9 seconds after ignition of the first stage main engines until the computer in the base, in constant communication with the firing room computer, verified that each engine had built up to full power. When the computer was satisfied, it released the hold-down arms and the rocket lifted from the launcher.

The launchers rest on 22-foot (seven-meter) tall pedestals in the VAB high bays or at the launch sites. A square opening in the launcher's base, directly under the Saturn V first stage, vented engine exhaust into a flame trench at the firing site.

The base of the mobile launcher was as much like a building as it was a platform for launching Saturn Vs. Within the launcher base were compartments, including a mechanical equipment room, operations support room, and television and communications equipment. Floors were suspended on springs, or shock insulators, and walls were lined with thermal and acoustical fiberglass insulation. Computers were housed in cocoonlike compartments enclosed by thick steel plates lined with fiberglass. Remotely operated, digitally controlled equipment in the base controlled propellant loading from the firing room.

During Apollo, two high-speed elevators, centrally placed in the umbilical tower, carried engineers and technicians to and from the swing arms and 17 work platforms. In an emergency the elevator used by the astronauts could be remotely controlled from the firing room, and the crew could reach the base of the mobile launcher in 30 seconds from the 320-foot (98-meter) level. There they could slide down a stainless steel chute, close to the elevator, which carried them four stories down inside the launch pad itself to a cavelike room lined with foam rubber to cushion their impact. Next to where the astronauts would land was a blast-resistant room containing 20 contour chairs and safety harnesses. The room had thick steel and concrete walls designed to withstand blast pressures of 500 pounds (227 kilograms) per square inch (6.45 centimeters) and an acceleration of 75 Gs. A floating concrete floor supported the contour chairs. It was built on a spring suspension system which reduced the 75 Gs possible on the room's dome to 4 Gs. Food and water were kept in the room for use in an emergency.

A second means of escape was also available to the Apollo astronauts. At the crew access arm level of the umbilical tower, a cage was suspended on a heavy cable. It was large enough to accommodate the three crewmen, plus the six technicians who assisted their entry into the spacecraft. The cage slid down the cable at 50 miles (80 kilometers) per hour, braking to a stop at an earth bunker 1,200 feet (366 meters) from the firing site. A 14-man rescue crew remained at the bunker

Astronauts test the emergency escape slidewire available for use during the Apollo program.

during final countdown and launch. This team could use armored personnel carriers to reach the pad quickly, or to haul the astronauts from the danger area.

Means of transporting the mobile launchers and assembled Apollo-Saturn V vehicles from the VAB to the launch pads were thoroughly investigated by KSC engineers. One mode involved a barge system, floating them from the VAB to the pads in a canal. Models were tested but found infeasible. A rail system was explored but proved too costly. Ideas for pneumatic-tired transporters, air-cushion vehicles, and others were explored and discarded.

The choice was a crawler rolling on eight tracks, its propulsion system resembling those used for large power shovels employed in strip mining In 1963 NASA awarded a contract for the construction of two crawlers. Both machines were in service by early 1967. Each weighs approximately six million pounds (2.7 million kilograms) and could transport a mobile launcher with the assembled Apollo-Saturn vehicle at a speed of one mile (1.6 kilometers) per hour. A trip to the Moon began very slowly. The crawler could also return the rocket to the VAB, if needed, with all connections intact.

In operation, the crawler moved underneath the mobile launcher and engaged its jacking system, which fitted under the platform. The launcher was raised off the support pedestals and carried into a VAB high bay, where it was carefully positioned on other pedestals. After the space vehicle had been assembled on the launcher and tested, the crawler returned to the high bay, again picked up the launcher, and carried it to the firing site.

The crawler-transporter, used for moving the assembled Apollo-Saturn V atop a mobile launcher to the pad, now carries the Space Shuttle atop its platform.

The combined mass of crawler, mobile launcher, and Saturn V added up to 18 million pounds (8,165,000 kilograms).

The roadway which supported this weight was different from a normal highway. Known as the crawlerway, it is about the width of the New Jersey Turnpike. Each of its two lanes is 40 feet (12 meters) wide with a 50-foot (15-meter) median. To prepare the base, fill was dredged from the adjacent Barge Canal, beginning in November 1963. The roadway was built in four layers, with an average depth of seven feet (two meters). The top surface on which the transporters operate is Alabama river gravel, a loose rock that relieves friction on the crawler treads. This roadway, completed in August 1965, has 10 times the resistance of a jet airport runway.

Lining the crawlerway along the north side are utility pipelines connecting the VAB and pads, while communications and instrumentation lines are buried in ducts below the surface. Midway between the Launch Control Center and Pad A, an extension of the crawlerway veers northeast and connects with the second launch site, Pad B.

Another major element of the mobile concept was the 10.5 million-pound (4.8 million-kilogram) mobile service structure. This 410-foot (125-meter) tall steel-trussed tower supported passenger and freight elevators, a power plant, air-conditioning equipment, a computer, and television and communications systems. Its function was to provide 360-degree access to the upper portion of the space vehicle for final launch preparations.

Assembled at its parking site along the crawlerway, the mobile service structure measured 135 feet (41 meters) by 132 feet (40 meters) at its base, or roughly half the size of a football gridiron, and was 113 feet (34 meters) square at its top. It supported five enclosed work platforms, two of which were self-propelled, designed to embrace the Apollo-Saturn V upper stages. The platforms enabled technicians to conduct final inspections, load propellants in the spacecraft, install ordnance items, and prepare the lunar, command, and service modules, and the launch escape system, for flight. (The escape system was a fingerlike rocket mounted atop Apollo which could develop sufficient thrust, in an emergency, to carry the spacecraft away from the launch vehicle. Then the Apollo command module would deploy its parachute and land in the ocean, where the crew could be rescued by amphibious vehicles or helicopters waiting in the launch area for this purpose.)

The structure was completed in late 1966. Since it could be moved, the same structure served vehicles at either pad. It was carried to the pad surface by the crawler after the mobile launcher and space vehicle were positioned over the flame trench. About seven hours before launch, at the time when liquid hydrogen propellant was to be loaded in the upper rocket stages, the crawler removed the mobile service structure from the pad and returned it to its parking position.

Pad A, the first of two launch pads to be constructed, was built between November 1963 and October 1965. The second firing site, Pad B, was identical. Both were ready for use by late 1968.

A critical item influencing pad design was the size and type of flame deflector required for the Saturn V vehicle. A flame deflector is a heavy steel device built in an inverted V-pattern which converts the vertical flames from booster engines into two horizontal flows along the trench. It was desirable to keep the colossal Saturn as close to the ground as possible, in order to avoid adverse wind effects at higher levels. Unless deflected, the steams of fire would bounce off the trench floor and back up along the vehicle.

Since the water table is close to the surface, the flame deflector had to be mounted at ground level. The flame trench in which it sits is lined with fire-resistant brick. It is 58 feet (18 meters) wide and 42 feet (13 meters) high. The launch pad's overall shape is roughly octagonal, and covers about one-half square mile (130 hectares). The flame trench and concrete hardstand is in the center of the fenced area. During launch preparations, the mobile launcher supporting the Saturn V vehicle rested on six 22-foot (6.7-meter) pedestals located on the pad surface. The other fixed facilities on the hardstand included a hydrogen service tower, a fuel system service tower, and six electrical power pedestals.

Beneath the surface were four floors. A terminal connection room housed electronic equipment which was integral to the communications system connecting the mobile launcher with the firing room. Other compartments accommodated environmental control systems, high-pressure gas storage systems, and the emergency egress room for the astronauts and spacecraft close-out crew. An independent water system was installed to cool the flame deflector following rocket ignition.

Around the pad perimeter were pressure storage tanks for RP-1, the kerosene fuel for the Saturn V first stage engines; liquid oxygen, at minus 297 degrees Fahrenheit (minus 183 degrees Celsius), the oxidizer for all three vehicle stages; and liquid hydrogen, at minus 423 degrees Fahrenheit (minus 253 degrees Celsius), which fueled the second and third stages. Holding ponds within the pad area retained spilled fuel and waste water. There was a burn pond for disposal of hydrogen gas boil-off. Stainless steel, vacuum-jacketed pipes carried liquid oxygen and hydrogen from storage tanks to the pad, then to the mobile launcher and ultimately to tanks inside the vehicle stages.

Fueling an Apollo-Saturn V space vehicle required several days of carefully coordinated operations. The spacecraft was first loaded with hypergolic propellants; that is, propellants that ignite upon contact with each other. Nitrogen tetroxide, the oxidizer, was loaded into the Apollo service module tanks, the lunar module ascent and descent tanks, and the reaction control system. Aerozine-50 fuel was then loaded into the service module and lunar module tanks. Monomethyl hydrazine and nitrogen tetroxide were loaded into the auxiliary propulsion system of the Saturn V third stage at the same time the spacecraft was fueled. This system of small thrusters operated during coast periods to provide enough velocity to keep the propellants settled in the bottom of their tanks.

When this loading was completed, the launch crews started piping propellants aboard the Saturn V. This operation was remotely controlled from the firing room. The computer measured the amount of propellant within the tanks by means of probes, while the computer within the mobile launcher base controlled flow rates by modulating loading valves within each stage interface.

The RP-1 fuel (kerosene) was taken aboard the Saturn first. Then, as the terminal count began on the third day, cryogenic (kept at extremely low temperatures) propellants (liquid oxygen and hydrogen) were pumped into the vehicle.

The new launch complex and the mobile concept showed up well during a stiff testing period in 1966. For this purpose, the Marshall Space Flight Center fabricated a mockup vehicle, designated Apollo-Saturn-500F (AS-500F), to check out launch facilities and train the rocket handling crews. It was precisely the shape and weight of the Saturn V and contained all the tankage, lines, electrical systems, and other components required to verify the launch facilities and equipment—except it had no engines.

On May 25, 1966, government and contractor employees gathered outside the VAB with astronauts to await the appearance of the gigantic AS-500F assembly. Dr. Robert C. Seamans, then deputy NASA administrator, spoke briefly and remarked crisply, "We are now going to see if the idea works." At his signal the crawler began to emerge from the bay. High overhead loomed the launcher tower; beside it was the gleaming white rocket—the largest ever seen in the United States—visible to the public for the first time just five years after President Kennedy announced the Apollo program.

The transfer to the pad went off smoothly. Weeks of arduous testing followed. NASA was pleased with the results—the facilities were ready on time.

From 1967 through 1973, there were 13 Saturn V launches from Complex 39. The first two were unmanned rocket and space flight development flights, and the next 10 carried three-man Apollo crews into space. The last Saturn V rocket boosted the unmanned Skylab space station into Earth orbit.

Complex 39 also served as the site for three manned Skylab launches by Saturn IBs in 1973, and for the manned Apollo-Soyuz Test Project flight in 1975—also boosted into Earth orbit by a Saturn IB.

During Apollo, 12 men explored the Moon. Skylab saw nine men spend extended periods of time living and working in an orbiting space station. During Apollo-Soyuz, American and Soviet crew members exchanged visits between their docked spacecraft in history's first international space mission.

Though Apollo-Saturn is history, Complex 39—its facilities modified and expanded to support Space Shuttle operations—continues as the nation's primary launch base for manned space flight ... a testament to the past and a steppingstone to a new era of exploration and utilization of the solar system for the benefit of all humanity.

chapter 6

DISASTER at COMPLEX 34

While facilities on Merritt Island were being prepared for the Apollo-Saturn V, testing continued on Cape Canaveral with the smaller Saturn I and IB vehicles, verifying the equipment and systems to be used on later Moon flights. The 13th Saturn flight—the third Saturn IB—on Aug. 25, 1966, proved to be the 13th success. It fulfilled all mission objectives. The team at KSC then began to prepare for the first manned Apollo mission at Launch Complex 34, scheduled for sometime late in 1966. The tragedy of Apollo-Saturn 204, later renamed Apollo 1, and its three-man crew would become etched in the annals of space flight.

NASA selected two veterans and one rookie as the crew for the first manned Apollo mission. Command Module Pilot Virgil Grissom had skippered Mercury's Liberty Bell 7, America's second suborbital manned flight, in July 1961, and Gemini's Molly Brown, the first manned Gemini, in March 1965. Edward White, while on the fourth Gemini flight, had become the first American to walk in space. With these two experienced space travelers would be the youngest American ever chosen to go into space, 31-year-old Roger Chaffee. The purpose of their flight was to check out launch operations, ground tracking and control facilities, and the performance of the command and service modules in orbit.

Following all the usual preliminary testing, the crew prepared for the simulated countdown test, the last major test scheduled prior to launch. There would be no fuel in the Saturn IB, designated Apollo-Saturn 204. Grissom, White, and Chaffee would don their full space suits and enter the Apollo, breathing pure oxygen to approximate orbital conditions as closely as possible. Apollo Mission Control at the Manned Spacecraft Center in Houston would pick up the action after simulated liftoff and would monitor the performance of the astronauts.

The astronauts entered the Apollo at Launch Complex 34 at 1 p.m., Friday, Jan. 27, 1967. Problems immediately arose. Grissom reported a strange odor in the suit loop which he described as a "sour smell somewhat like buttermilk." After taking a sample of the suit loop, the crew decided to continue the test. The next problem was a high oxygen flow indication which periodically triggered the master alarm. The high flow was believed to be the result of movements made by the crew. A third serious problem arose in communications, first between the control room and Grissom and then later extending between the blockhouse at Complex 34 and other spacecraft monitoring facilities. The communications problem forced a hold of the countdown at 5:40 p.m. By 6:31 p.m., the test conductors were preparing to resume the count when ground instruments showed an unexplained rise in the oxygen flow into the space suits of the crew. One of the crew, presumably Grissom, moved slightly.

Four seconds later, an astronaut, probably Chaffee, announced almost casually over the intercom, "Fire, I smell fire." Two seconds later, astronaut White's voice was more insistent, "Fire in the cockpit!"

Inside the blockhouse, engineers and technicians looked up from their consoles to the television monitors trained at the spacecraft. To their horror, they saw flames licking furiously inside the smoke-filled Apollo. Men who had gone through Mercury and Gemini tests and launches without a major incident stood momentarily stunned.

Outside the white room that gave access to the spacecraft, emergency procedures to rescue the astronauts were ordered. Technicians started toward the spacecraft. Then the command module ruptured. Flames and thick black smoke billowed out, filling the room. A new danger arose. The fire and heat might set off the launch escape rocket atop Apollo. This, in turn, could ignite the entire service structure. Instinct told the technicians to get out while they could. Many did, but six remained and continued rescue attempts. Despite the intense heat, thick smoke and the danger overhead from the escape rocket, they managed to get Apollo's hatch open. But it was too late. The astronauts were dead. A medical board determined later that the crew died of carbon monoxide asphyxia, with thermal burns as a contributing cause.

The sudden and unexpected deaths of the three astronauts caused international grief and widespread questioning of the space effort. Momentarily, the whole manned lunar program stood in suspense.

On Feb. 3, NASA Administrator James Webb set up a review board to perform a thorough investigation. That same day, ground crews at KSC began to sift through the burned hulk of Apollo 204. At the Manned Spacecraft Center in Houston, engineers duplicated conditions of Apollo 204 (without crewmen in the capsule), reconstructing events as studies at KSC brought them to light.

The investigation at Pad 34 showed that the fire started in or near one of the wire bundles to the left and just in front of Grissom's seat on the left side of the cabin—a spot visible to Chaffee. The fire probably was invisible for about five or six seconds, until Chaffee sounded the alarm. From then on, the test fire duplicated by the Manned Spacecraft Center engineers followed, almost to a second, the pattern of intensity of the oxygen fire aboard Apollo 204.

On April 5, the review board submitted its formal report to the administrator. It summarized its findings as to the cause of the fire in this sequence: "The fire in Apollo's floor was most probably brought about by some minor malfunction or failure of equipment or wire insulation. This failure, which most likely will never be positively identified, initiated a sequence of events that culminated in the conflagration."

Dr. Debus, KSC's director, summed up the feelings of the launch team in a statement before a congressional hearing. Speaking candidly, he said: "As director of the installation, I share the responsibility for this tragic accident and I have given it much thought. It is for me very difficult to find out why we did not think deeply enough or were not inventive enough to identify this as a very hazardous test ...

From left to right, astronauts Virgil Grissom, Edward White and Roger Chaffee. Out of the analysis and soul-searching that followed their tragic deaths, America's Apollo program was rededicated.

"We never knew that the conflagration would go that fast through the spacecraft so that no rescue would essentially help. This was not known. This is the essential cause of the tragedy. Had we known, we would have prepared with as adequate support as humanly possible for egress."

NASA moved quickly to better assure that a like tragedy would not occur in the future. A new flameproof material called Beta Cloth was substituted for nylon in the space suits. Within the spacecraft itself, technicians covered exposed wires and plumbing to preclude inadvertent contact. They redesigned wire bundles and harness routings. The cabin atmosphere was changed from 100 percent oxygen to 60 percent oxygen and 40 percent nitrogen.

At Complex 34 itself, technicians put a fan in the white room for ventilation. They added water hoses, fire extinguishers, and an escape slidewire. Astronauts and crew workers could ride down this wire during emergencies, reaching the ground in seconds.

Top priority was given to redesigning the hatch. The new Apollo hatch would be a single hinged door that swung outward with only one-half pound (0.23 kilograms) of force. An astronaut could unlatch the door in three seconds. The hatch would have a push-pull unlatching handle, a window for visibility in flight, a plunger handle inside the command module to unlatch a segment of the protective cover, a pull loop that permitted a pad man to unlatch the protective cover from the outside, and a counterbalance to hold the door open.

Then, after the most careful consideration, a difficult decision was made: After installation of new equipment and modifications of procedures, the Apollo program, with its objective of landing on the lunar surface by the end of the decade, would continue.

chapter 7

STEPS TOWARD LUNAR LANDING

For the men and women who planned the Apollo program in 1961, and for thousands of others who designed, fabricated, and tested the Saturn V launch vehicle and Apollo spacecraft, the morning of Nov. 9, 1967, was to be the ultimate test. The years spent in planning and constructing facilities, and in organizing the largest launch team ever assembled in this country, were about to pay off.

Since the earlier Saturns had been designated the 200 series, NASA had numbered the Saturn Vs the 500 series. Consequently, the first to be launched became Apollo-Saturn 501, and since this was the fourth Apollo scheduled, the mission became Apollo 4.

The flight hardware coming off production lines and static test stands had been arriving at the Center since mid-1966. At nightfall on Aug. 25, 1967, the crawler moved into the VAB, jacked up the mobile launcher and space vehicle and prepared to transfer the 12 million-pound (5.4 million-kilogram) assembly to the launch pad. Early in the morning of Aug. 26, the towering mass of the mobile launcher carrying Apollo 4 slowly emerged from the building.

For the new explorers of a new frontier, movers and planners such as Dr. Wernher von Braun, left, and Dr. Kurt Debus, the Apollo-Saturn program would be the realization of a dream.

In mid-September, bad weather hampered tests of the liquid hydrogen fueling system at the pad. Gale-force winds were measured on the mobile launcher. The following week, lightning and heavy rainfall postponed the countdown demonstration test, which verifies that rocket and spacecraft systems and equipment are ready for launch.

The first Apollo mission for Saturn V, the new and untried giant booster, began at a snail's pace with rollout from the Vehicle Assembly Building to the launch pad at Complex 39.

Three times between Sept. 27 and Oct. 31, the test was temporarily halted by a succession of annoying problems. When it was finally completed, Center management felt that a major step had been accomplished—the approximately 450-person crew in Firing Room 1 and the men on the pad had coordinated their efforts over a three-shift working day, developing confidence among the government/industry team members.

The 104-hour countdown began Oct. 30 when the spacecraft was fueled, and RP-1 kerosene fuel was loaded into the first stage. On Nov. 8, wind velocities were marginal. The Apollo-Saturn V configuration could not be launched if the steady wind force was at 28 knots (52 kilometers). The weather forecaster predicted that maximum gusts would not exceed 26 knots (48 kilometers) at scheduled liftoff time on Nov. 9. It was decided to proceed with the terminal count and look at the wind situation again later.

Preplanned holds were provided in the countdown. Some of this reserve time was consumed when a scratched seal had to be replaced in the Saturn V. Another two-hour hold became necessary to check out the range safety command system.

The count proceeded until 3 a.m., Nov. 9, when the clock was stopped for 60 minutes in order to reach T minus zero at 7 a.m. Then there would be ample light for cameras to record every event on the launch pad, as well as the behavior of the launch vehicle from ignition until it disappeared out of camera view downrange.

Ignition occurred exactly on schedule, 8.9 seconds before lift-off. Six seconds later the giant engines had built up to 90 percent thrust. The hold-down arms constraining the rocket to the deck released at 7:00:01 a.m. EDT. The first Saturn V had begun its journey.

Following burnout and separation of the first and seconds stages, the third stage flamed into life to insert the spacecraft into Earth orbit. The stage was shut down, then later re-ignited to increase both the speed and altitude of the combined vehicle. When the third stage burned out, it was jettisoned. Then the Apollo spacecraft propulsion system burned for 16 seconds, raising its altitude to over 11,170 miles (18,000 kilometers). Apollo was reoriented and the engine fired once more to drive the spacecraft back into the atmosphere at 25,000 miles (40,200 kilometers) per hour, the speed Apollo would reach on its return trip from the Moon.

Ignition--and the first Saturn V flight lifts Apollo a step closer to the Moon.

After spacecraft recovery, examination showed that the heat protection afforded by the blunt ablation shield on the command module had withstood the test of more than 5,000 degrees Fahrenheit (2,760 degrees Celsius). The temperature inside the Apollo spacecraft climbed only 10 degrees, and never exceeded 70 degrees. Performance of both rocket and spacecraft had been letter-perfect throughout the mission.

Following the success of Apollo 4, preparations moved ahead smoothly at Complex 37 for the first orbital test of the lunar module—the buglike spacecraft designed to land men on the Moon. The launch vehicle would be a Saturn IB, Apollo-Saturn 204.

Known as Apollo 5, the mission began the morning of Jan. 22, 1968, when the unmanned rocket rose from its launch pad carrying the lunar module. Within minutes, the spacecraft was inserted into Earth orbit as planned, then separated from its carrier rocket.

From that point on, the module responded both to programmed commands and to new instructions transmitted by radio links from the Mission Control Center in Houston. Both ascent and descent propulsion systems operated satisfactorily. The flight demonstrated that these systems could be throttled while operating, and that the engines could be started, stopped and restarted in space. No attempt was made to return the lunar module to Earth. It was not designed to withstand re-entry heating and could only function properly in space or in the one-sixth Earth gravity on the surface of the Moon.

On Feb. 6, the second Saturn V launch vehicle was carried by the crawler-transporter from the VAB to Pad A in preparation for the launch of Apollo 6. This was to be another test of the launch vehicle, and the second flight test of the command module's ability to withstand re-entry heating at lunar return velocities.

Apollo 6 was launched on schedule the morning of April 4, carrying a payload of 93,885 pounds (42,586 kilograms) into Earth orbit. The countdown went smoothly. Both vehicle and spacecraft responded to all checkout procedures and prelaunch tests. However, some problems occurred during the powered flight of the Saturn V first stage, including severe up-and-down vibrations. Then one of the five engines in the hydrogen-fueled second stage shut down prematurely, and three seconds later a second engine ceased to function. To compensate for this loss of thrust, the third stage burned 29 seconds longer than planned. This resulted in the Apollo spacecraft being inserted into orbit, and the mission continued.

Later, another problem developed. The third stage failed to re-ignite as programmed. Consequently, it did not boost the Apollo spacecraft high enough above the Earth to simulate a return from the Moon. After Apollo had been separated from the stage, it was maneuvered higher by ground control of the service module propulsion system. From this elevation Apollo was steered into re-entry almost at the original planned speed. The 10-hour flight terminated with the recovery of the spacecraft in the Pacific—Apollo had stood up well under this severe test.

As the launch vehicle designers examined the performance data accumulated during the flight of the Saturn V, they determined that the shutdown of the second stage engines had occurred because of a wiring error. They also found that spark igniters linked to

Awakened at 4:15 a.m. on July 16, the astronauts breakfasted on orange juice, steak, scrambled eggs, toast and coffee, then began suiting up at 5:35 a.m. They departed for the pad at 6:28 a.m., arriving after fueling operations had been completed. Armstrong entered the Apollo at 6:45 a.m., assisted by the six-man pad close-out crew under the direction of Rockwell International's Guenter Wendt and NASA's Spacecraft Test Conductor "Skip" Chauvin. Collins joined Armstrong five minutes later, sliding into the right couch. He was followed by Aldrin, who climbed into the center seat. Before leaving the pad at 8:32 a.m., the close-out crew shut the hatch, pressurized the cabin to check for leaks, and purged it.

Apollo 11 astronauts Neil Armstrong, Edwin Aldrin Jr., and Michael Collins leave the Manned Spacecraft Operations Building, bound for a unique page in the history of flight.

Three and a half miles (5.6 kilometers) away, in the Launch Control Center, the 463-person Apollo launch team monitored the final minutes of the countdown. Launch Director Rocco Petrone directed the operation from his command station in "management decision row" which looked out over row upon row of flashing consoles and recording racks manned by engineers and technicians from the space agency and its major Apollo contractors—Boeing, Rockwell International, McDonnell Douglas, IBM, General Electric, Grumman, Rocketdyne and Aerojet General. Seated alongside Petrone were Dr. Debus, KSC's director; Dr. Hans Gruene, the Center's director of Saturn V operations; and Dr. George

Low, Apollo program manager for the Johnson Space Center in Houston. Observing the countdown procedures were NASA Administrator Dr. Thomas O. Paine and other high-level NASA officials, including Dr. George Mueller, associate NASA administrator; Lt. Gen. Samuel Phillips, NASA's Apollo program chief; and Dr. Wernher von Braun, father of the Saturn V rocket and director of the Marshall Space Flight Center in Huntsville, Ala.

On the first floor of the Launch Control Center, technical support personnel manned consoles monitoring propellant flow, life support, and other ground systems.

Five miles (eight kilometers) south in the Industrial Area, two Acceptance Checkout Equipment stations manned by General Electric Co. personnel in the Manned Spacecraft Operations Building monitored Columbia and Eagle.

Engineers in the nearby Central Instrumentation Facility recorded measurements from the Saturn V and monitored the intricate communications network tying together all the operating stations.

In the Launch Control Center firing room 3.5 miles (5.6 kilometers) west of the launch pad, the 463-person Apollo 11 launch team monitored the last suspenseful moments of the countdown.

Over the past dozen hours or so, the launch team had checked off thousands of items from the three-inch (7.6-centimeter) thick countdown manual, the bible of Saturn launch operations. The huge launch status board showed all green. Two minor equipment problems, a leaky valve and a faulty signal light, had been corrected while the astronauts were en route to the pad.

Wednesday, July 16, was a beautiful morning—bright sunshine, a few fleecy clouds and a slight wind from the southeast. By 9 a.m. it was very warm. Five thousand guests were gathered at the viewing site north of the VAB. Among them were Vice President and Mrs. Spiro Agnew, former President and Mrs. Lyndon B. Johnson, Army Chief of Staff Gen. William Westmoreland, Chief Justice of the U.S. Supreme Court Earl Warren, the Archbishop of New York Cardinal Terence Cooke, aviation pioneer Charles Lindbergh, writer William F. Buckley, television personalities Johnny Carson and Ed McMahon, and comedian Jack Benny. Also in the crowd were 33 U.S. senators, 206 congressmen, 56 ambassadors, and the secretaries of health, education and welfare; commerce; transportation; and interior.

Close by was the press corps, more than 1,700 strong, representing 56 nations, including three Communist countries. Thousands of NASA civil service and contractor employees and their families lined the perimeter roads of the complex. And miles away, jamming the riverfronts, beaches and highway approaches to the Space Center, were an estimated one million visitors, possibly the largest crowd in history to witness a single launch.

About 2,600 feet (792 meters) from the launch pad, protected by a sand bunker, 14 rescue personnel stood watch as usual for manned Apollo launches. Equipped with armored personnel carriers and wearing flame protection gear, they could quickly assist the astronauts in an emergency. Manning special roadblocks surrounding the complex were teams of doctors, nurses, safety officials, ordnance experts and recovery specialists—prepared to spring into action if the Saturn V exploded on the pad. Loaded with nearly 3,000 tons (2,722 metric tons) of volatile fuel, an exploding Moon rocket would shatter with a force equivalent to 1.2 million pounds (544,300 kilograms) of TNT.

High in the Apollo command module, the astronauts gripped the armrests of their couches, ready, if necessary, to trigger the launch escape tower atop Apollo which would pull them free of the rocket and parachute them back to the ground some distance from the pad. Firemen and medical specialists would speed to the scene in moments aboard helicopters and amphibious vehicles.

The Apollo access arm was retracted at 9:27 a.m., or T minus 5 minutes in the count. At 4 minutes the "cleared for launch" command was given. The countdown became automatic at 3 minutes, 20 seconds. Apollo 11 lifted off at 9:32 a.m. EDT.

Apollo 11 attained Earth orbit 11 minutes and 49 seconds after liftoff. Cleared to proceed to the Moon, the astronauts fired the third stage engine again at 12:22 p.m., increasing the velocity to 24,000 miles (38,600 kilometers) per hour. Collins then separated Columbia, turned it around and docked with Eagle. During the third day, Armstrong and Aldrin removed the docking probe and drogue and opened the tunnel hatch. They entered the Eagle to perform housekeeping chores and check the equipment. Their activities

Liftoff for Apollo 11--and the voyage of exploration on a vast new sea is begun.

were seen on television by millions of people in the United States, Japan, South America, Canada, and Western Europe. The same day, the astronauts entered lunar orbit.

On July 20, Armstrong and Aldrin again occupied Eagle, powered it up and deployed its landing legs. Eagle and Columbia separated at 1:46 p.m. Armstrong and Aldrin fired Eagle's descent engine at 3:08 p.m. Forty minutes later, as Columbia emerged from behind the Moon, Collins reported what had occurred, commenting, "Everything's going just swimmingly."

As the moonscape came into clearer view, Armstrong saw that Eagle was approaching a crater almost as large as a football field. He took over manual control and steered toward a less formidable site. At Mission Control, physicians noted his heartbeat increasing from a normal 77 to 156 beats per minute. While Armstrong manipulated the controls, Aldrin called out altitude readings, ending with "... contact light ... O.K. Engine stop." As the probes beneath three of Eagle's four footpads touched the surface, a light flashed on the instrument panel. The world heard Armstrong's quiet message.

"Houston, Tranquility Base here. The Eagle has landed."

Later the crew explained that while some distance from the surface, fine dust blew up around the spacecraft and obscured their vision. However, they felt no sensation at the moment of landing.

"The Eagle has landed" . . . and a "giant leap for mankind" leaves its print on the Moon.

At 6 p.m., Armstrong called Mission Control to recommend that the walk on the Moon begin about 9 p.m., earlier than originally planned. However, it was 10:39 when Armstrong opened the hatch and squeezed through it—still five hours ahead of schedule. He wore 84 pounds (38 kilograms) of equipment on his back, containing his portable life support and communications systems. On the Moon the weight was 14 pounds (6.35 kilograms).

Armstrong proceeded cautiously down the nine-step ladder. He paused at the second step to pull a ring which deployed a television camera, mounted to follow his movements as he climbed down. At 10:56 p.m. Armstrong planted his left foot on the Moon, saying as he did: "That's one small step for a man, one giant leap for Mankind."

Later as he described the powdery lunar surface material and collected soil samples, he remarked, "It has a beauty all its own. It's like much of the high desert in the United States."

Aldrin emerged from Eagle and joined Armstrong at 11:11 p.m. For the next two hours they collected rock samples, set up scientific apparatus, erected an American flag, took pictures and loped easily across the surface while an estimated 600 million Earth viewers watched via television.

Having completed their assigned tasks, the astronauts re-entered Eagle on instructions from Mission Control, closing the hatch at 1:11 a.m., July 21. They tried to rest but could not. The lunar module was cold and noisy. At 1:45 p.m., July 21, having spent

As Apollo 11 astronauts spent two hours on the surface of the Moon, an estimated 600 million people on Earth shared the adventure via their television sets.

Earth rises over the lunar horizon, viewed from orbit around the Moon. In the end, Apollo's lessons will be applied to protecting and preserving "Spaceship Earth."

22 hours on the lunar surface, Aldrin counted down and fired the ascent stage engine, which functioned perfectly. They docked with Columbia and rejoined Collins at 5:35 p.m. Collins touched off the Apollo main engine at 12:55 a.m., July 22, while in the Moon's dark side. Columbia headed home. The spacecraft's splashdown site was changed due to stormy seas. The astronauts adjusted their course to a new site 270 miles (435 kilometers) away, hitting the Pacific Ocean at 12:50 p.m. EDT, July 24. President Richard M. Nixon and NASA Administrator Paine were on the aircraft carrier to greet them.

The astronauts climbed into a mobile isolation trailer aboard ship. They remained inside the trailer while being transported to Houston by ship, airplane and truck. After their arrival in Houston, they entered quarantine in the Lunar Receiving Laboratory as a safeguard against bringing any possible hostile organisms back to Earth. Quarantine ended Aug. 12.

The Apollo program had achieved its objective five months and 10 days before the end of the decade.

To the Moon ... and the loftiest objective in history is achieved, five months and 10 days before the end of the decade.

chapter 9

APOLLO CONTINUES

NASA selected a veteran space pilot, Charles "Pete" Conrad, to command Apollo 12, the second lunar landing attempt. Richard Gordon became the command module pilot, and Alan Bean filled out the crew as lunar module pilot. Bean had not previously flown in space.

The crew took up residence at KSC in mid-August 1969 for final preflight training. Prior to the crew's arrival at KSC, the launch team had worked on the rocket and the

Apollo 12 astronauts Charles Conrad, left, and Alan Bean train for planned activities in the Moon's Ocean of Storms, during a simulation staged in the Flight Crew Training Building.

spacecraft. The first of the rocket stages had arrived March 9, six days after the Apollo 9 launch. Ascent and descent stages of "Intrepid," as the all-Navy crew dubbed the lunar module, arrived later that month. So did the command and service modules, named "Yankee Clipper."

With September came a change in KSC management. Petrone left the launch director's post to become the Apollo Program director at NASA Headquarters in Washington. His deputy, Walter J. Kapryan, succeeded him.

Apollo 12 rolled out of the Vehicle Assembly Building en route to the pad at daybreak on Sept. 8. The launch countdown had begun at 8 a.m. Nov. 7, and the clock had started at T minus 108 hours. The astronauts passed their final physicals on Nov. 10.

A problem developed at T minus 40 hours. One of two hydrogen tanks in Yankee Clipper's service module failed to chill down when the extremely cold liquid propellant was pumped aboard. Tanking continued until both tanks were 90 percent full. The quantity in tank number 2 continued to drop and frost formed on the outer shell.

This was interpreted to indicate either that the inner shell was leaking, allowing hydrogen to flow between the shells, or that a leak had occurred in the outer shell. The decision was made to remove the suspect tank and replace it with a tank from Apollo 13. Technicians worked around the clock to make the substitution and the countdown proceeded. The launch team also installed a small atomic generator fueled by plutonium-238 which would provide power for the experiments the crew would leave on the lunar surface.

The crew entered Apollo 12 on the morning of Nov. 14. The weather had turned unpleasant during the night and rain fell intermittently. Dark clouds moved northeasterly across the launch site. Launch Director Kapryan kept in close touch with the KSC weather station, whose readings were supplemented by two aircraft flying through and just above the clouds over the spaceport. No lightning was reported. Heavy rain fell about an hour before launch. President and Mrs. Richard Nixon arrived 40 minutes before liftoff, joining other observers at a viewing site north of the Vehicle Assembly Building. Apollo 12 lifted off precisely on schedule at 11:22 a.m. EST.

At the 36-second mark into the flight, spectators saw two flashes of lightning streak to the ground on either side of the launch tower. Conrad reported: "We just lost the platform. I don't know what happened here. We had everything in the world drop out ..."

Conrad was referring to the inertial platform that is the heart of the Saturn V navigation, guidance and control system. Apollo 12 never faltered. Kapryan later compared the power outage to a fuse blowout in a home, saying it protected vital electronic instrumentation from power overloads. The crew re-established the inertial platform and soon had power flowing in all systems.

Later, it was theorized that the vehicle plus its long flaming tail acted like a lightning rod, triggering static electricity in the cloud cover into lightning bolts.

After this incident, the mission became a textbook flight. On Nov. 19, having separated from Yankee Clipper, Intrepid entered a looping orbit which carried it down to the Ocean of Storms, landing within 600 feet (183 meters) of Surveyor 3, an unmanned lunar spacecraft, which had landed on the Moon two years before.

In all, Conrad and Bean spent 7 hours, 45 minutes working on the Moon, setting up scientific experiments, collecting lunar samples and pieces from the Surveyor, and photographing their landing craft and other objects of interest. They lifted off in the Intrepid ascent stage at 9:25 a.m., Nov. 20, and rendezvoused with the Yankee Clipper piloted by Dick Gordon. Clipper splashed down in the South Pacific at 3:58 p.m. on Nov. 24.

After quarantine the crew's first stop was KSC. When they returned Dec. 17, over 8,000 spaceport employees gathered inside the VAB to welcome them "home."

"The crew didn't consider the flight over until we got back here," Conrad told them, adding, "I'd just like to tell you, you all did a hell of a job for us."

World interest in the lunar landing program waned after the routine and highly successful Apollo 12 mission. It was dramatically rekindled by the plight of Apollo 13.

For the Apollo 13 mission, NASA selected space veteran James A. Lovell Jr., as commander, Thomas K. Mattingly II, command module pilot, and Fred W. Haise Jr., lunar module pilot. Mattingly and Haise would be making their debuts in space.

KSC began preparations for the mission in June 1969, when the first launch vehicle stage arrived. The astronauts named the command service module "Odyssey" and the lunar module "Aquarius." Their target on the Moon was Fra Mauro, a hilly area of major interest to scientists.

The rocket with the spacecraft atop was moved to Pad A in December 1969. From the beginning, the mission seemed jinxed. On March 25, 1970, the last day of the countdown demonstration test, a strange accident occurred. A large quantity of liquid oxygen used to chill down the liquid oxygen pumping system on the booster stage was emptied into a drainage ditch outside the pad perimeter fence—a routine procedure during tests.

Normally, ocean breezes dissipate the oxygen. However, on this morning there was a pronounced temperature inversion and no wind. A dense oxygen fog built up in the drainage ditch and overflowed onto a nearby roadway. A three-car security team, which had cleared the pad area, had stopped nearby. When one of the guards turned his ignition on, he heard a loud pop and flames sprang from beneath his hood. In rapid succession, the other two cars burst into flames. The three guards ran for cover. It was nearly an hour before the oxygen cloud dissipated and the fire could be brought under control.

The incident, which left three burned-out cars and a shaken security team, proved once again that ground operations were as fraught with unknown dangers as flying in space. KSC officials took immediate steps to eliminate the problem by changing operational and safety procedures and extending the liquid oxygen drainage pipes beyond the perimeter ditch into a marshy area further from the pad.

Another problem that arose during the test appeared insignificant at the time but, in fact, was the beginning of what were to prove Apollo 13's most nerve-wracking hours. The number 2 liquid oxygen tank in the service module, one of two liquid oxygen tanks that feed the fuel cells which supply electrical power and life support systems on the Apollo, failed to empty completely during repeated tests. Only by energizing the tank's heater and venting the tank were crews able to empty its contents.

The problem was thought to be a loose filler tube in the tank. Replacement of the tank, however, would take two days and posed the possibility of damaging other vital equipment. Moreover, a loose filler tube would not threaten the mission since it had no effect on the flow of oxygen to the fuel cells.

After studying the problem, Apollo officials at KSC, Houston, Washington and the manufacturer's plant in Downey, Calif., decided to keep the defective tank. On April 5, the final countdown to launch was initiated. Two days later, the Apollo 13 jinx struck the flight crew.

Astronaut Charles M. Duke, Jr., a member of the backup crew, became ill with the German measles, or rubella. John Young and John L. Swigert Jr., who trained with him, had been exposed to possible contagion, as had the prime crew. Following tests, Dr. Berry announced that Lovell and Haise showed immunity to rubella, but Mattingly did not. Swigert, the backup command module pilot, was also immune.

NASA reviewed the alternatives. Delaying the launch would be costly. On April 10, it was announced that Swigert would replace Mattingly because it would be unwise to risk the possibility that the command module pilot might develop measles during the mission, particularly when he would pilot Odyssey around the Moon alone while his crewmates were on the lunar surface.

The terminal count began at 4:13 a.m., April 11. Liftoff occurred on schedule at 2:13 p.m. During the ascent phase, the center engine of the Saturn V's second stage cut off more than two minutes early, and, to compensate, the remaining four engines were burned 34 seconds longer than planned. As a further remedy, the engine of the third stage was fired an extra nine seconds during its orbital insertion burn.

Despite these minor problems, the early events of the flight proceeded with gratifying smoothness. Fifty-five hours into the mission, the crew entered the lunar module Aquarius. A telecast fromspace followed, lasting about 30 minutes. Then, disaster struck. All three astronauts heard a loud bang. Swigert felt the spacecraft vibrate. Within two seconds, the master alarm sounded.

Mission Control was stunned by the terse words flowing over the radio link. "... Houston, we've got a problem here!" said the voice from space.

The nature and dimensions of the problem quickly became evident to the crew and Mission Control. Liquid oxygen tank number 2 in the service module—the tank found defective during ground tests—had exploded, wiping out the fuel cells that supplied life-sustaining oxygen and electrical power for the command and service modules. There was a backup battery-powered electric supply in the spacecraft, but, under ideal circumstances, it had a lifetime of only 10 hours.

Lovell and his crewmates were nearly 240,000 miles (386,243 kilometers) out in space and 87 hours from home. The service module, including the main propulsion engine which was needed to get them out of lunar orbit and on the way home, was dead. The command module's 10-hour battery supply had to be reserved for the approach to the Earth's atmosphere, for the command ship alone carried the vital heat shield and parachutes for safe re-entry and splashdown.

Now that the mother ship was a partial wreck, the crew's hope for salvation rested with the life support systems of the lunar module—the spiderlike craft designed to accommodate just two astronauts, not three. Still linked to its crippled parent, Aquarius had to become a lifeboat in space.

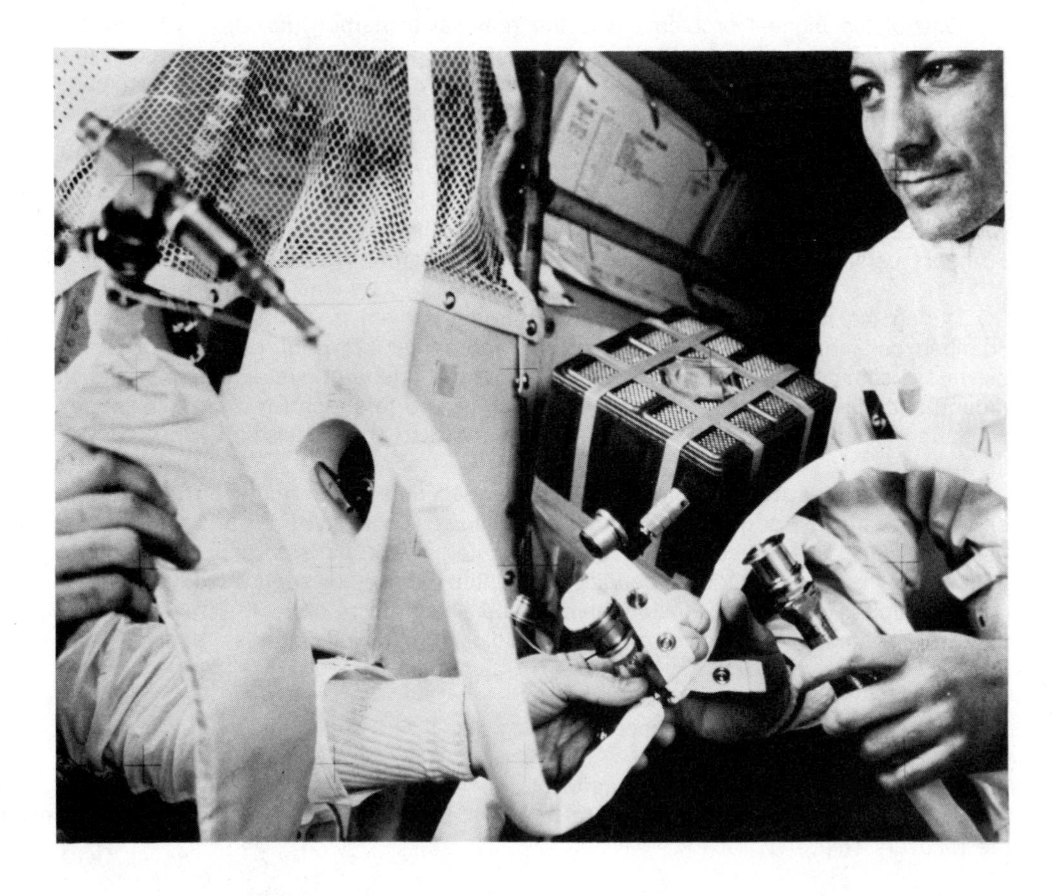

Apollo 13 Command Module Pilot John Swigert connects hoses allowing use of command module lithium hydroxide to scrub the carbon-dioxide-laden lunar module atmosphere.

Two major activities dominated the remainder of the mission: planning and conducting the mandatory propulsion maneuvers, utilizing the lunar module's descent engine as a substitute for the defunct service module engine; and managing the vital resources—oxygen, water, electricity, and the canisters of lithium hydroxide used to remove carbon dioxide from the cabin atmosphere—in the two spacecraft. Open communication lines between KSC and Mission Control in Houston carried advice and test requirements. The two Centers simulated the various maneuvers and conservation measures before directions were given to the flight crew.

A KSC team devised a means of recharging the command module's re-entry batteries from the lunar module's electrical system. Another KSC recommendation turned off the radar heaters to save electricity. Rockwell and Grumman engineers at KSC helped devise ways to transfer water from the portable life support systems designed for lunar surface activity into the lunar module's water coolant system.

One of the biggest problems was the removal of carbon dioxide from crowded Aquarius. KSC engineers, again duplicating activities at Houston, rigged a system that carried carbon dioxide-rich air from the lunar module through a hose into the command module's lithium hydroxide canisters. When the flight crew duplicated the procedure, carbon dioxide in the cabin immediately returned to tolerable levels.

The lunar module's descent engines performed beyond expectations, looping the two spacecraft around the Moon and into a trajectory that would bring them back to Earth 30 hours earlier than expected, and to the original target area in the South Pacific.

During the hectic voyage home, the astronauts lived in Aquarius, preferring its cramped confines to the chilly 52 degrees Fahrenheit (11 degrees Celsius) of the powerless command module Odyssey. Respect for the lunar module increased among ground and flight crews alike as its systems, designed to support two crew members, continued to sustain the three-man crew well past its two-day mission expectancy.

On April 17, the crew re-entered the command module and switched on its batteries, employing a phased power-up sequence to conserve electricity. Four and a half hours before re-entry, they jettisoned the service module and maneuvered their spacecraft to photograph the service module's condition. An entire panel of the service module housing had been ripped off by the explosion. An hour and a half from re-entry, the crew jettisoned their lunar module lifeboat. Mission Control radioed, "Farewell Aquarius, and we thank you."

Lovell added, "She was a good ship."

Odyssey re-entered the Earth's atmosphere 142 hours, 40 minutes and 47 seconds after the flight began, splashing down at 1:08 p.m. EST, 3.5 miles (5.6 kilometers) from the recovery ship.

NASA Administrator Dr. Thomas O. Paine ordered a Board of Review to look into the causes of the mishap and recommend corrective actions. After an inquiry, the board found that two thermostatic switches, which controlled electrical feed to heaters in the service module's number 2 oxygen tank, probably welded permanently in a closed position when activated during the countdown demonstration test at KSC in March. Switch failure was blamed on a change of manufacturer specifications in voltage to the switch—a change which somehow went unnoted over a period of time. Because of the inoperative switch, temperature and pressure rose in the tank when activated during flight; wiring insulation apparently ignited and more pressure was created. The result was a fire and rupturing of the tank. The review board recommended that all potential ignition sources and combustible materials be removed from oxygen tanks in the future. These precautions were taken before Apollo 14 left Earth.

When the Apollo 13 flight crew returned to KSC to speak to some 7,000 employees in the VAB, they presented the Center with an armrest from Aquarius as a permanent token of appreciation. It had been removed from the lunar module before the module was jettisoned.

Astronaut Lovell told the assemblage, "We're proud to come back today and tell you, `thank you.' I think the mission matured the space program a little, because people were perhaps getting a bit complacent about what we do."

The crew for Apollo 14 was announced in August 1969. America's first astronaut, Alan B. Shepard Jr., was named commander of the flight.

His crewmates would be Lunar Module Pilot Edgar D. Mitchell and Command Module Pilot Stuart A. Roosa. Both had joined NASA in 1966 and were experienced research pilots. It would be the first trip into space for both.

The launch was first targeted for October 1970, with the Littrow region of the Moon as the destination. Apollo 13's difficulties, however, forced changes in the flight plan as well as modifications to the spacecraft. The launch was postponed, first to December 1970 and then to Jan. 31, 1971—the 13th anniversary of the launch of the first U.S. satellite, Explorer 1. The rugged Fra Mauro highlands, Apollo 13's intended target, became the crew's new destination.

While the astronauts devoted their time to rigorous training, launch crews at KSC modified the spacecraft modules, checked them out, and tested and assembled the Saturn V stages. A Saturn V rocket had more than three million parts. Of particular concern to KSC engineers was the center engine of the rocket's second stage. During Apollo 13's flight, the second stage had experienced severe oscillations, known as the pogo effect, forcing an early shutdown of the center engine. Although the stage's remaining engines had burned longer to compensate for the loss, NASA officials did not want any more pogo. Engineers made several changes in the stage's engine systems to dampen the oscillations, including a cutoff device to shut down the center engine in case the other changes failed to correct the problem.

Apollo 14 was moved from the Vehicle Assembly Building to the launch site on Nov. 9, 1970. The countdown proceeded uneventfully until the sky clouded over and rain fell. Two aircraft carried instrumentation which measured electrical fields in the clouds and fed data into the Launch Control Center. Launch Director Walter Kapryan, following procedures established after the Apollo 12 lightning incident, called a hold around eight minutes before ignition to await more favorable weather. It was the first time an Apollo launch had been delayed. When the aircraft indicated danger from lightning had passed, Kapryan resumed the count. Apollo 14 lifted off at 4:03 p.m. EST, less than an hour behind schedule.

Three and a half days later the crew prepared for lunar descent. This was accomplished through a maneuver quite different from that executed by either Apollo 11 or 12. For those missions, the command ship ventured no closer than 70 miles (113 kilometers) from the surface and the lunar module separated at that altitude to begin the descent. The hilly terrain of Fra Mauro called for a steeper descent for the Apollo 14 lunar module, "Antares." The command service module, "Kitty Hawk," therefore approached the Moon within 10 miles (16 kilometers).

Antares landed on a slope of 18 degrees, just 87 feet (27 meters) north of the chosen site. The touchdown occurred at 4:18 a.m. EST on Feb. 5. After rest and preparations, the two men opened the hatch, climbed down to the surface and set up the Apollo Lunar Surface Experiments Package (ALSEP). The crew deployed a two-wheeled cart and began collecting samples, with Shepard describing the terrain samples.

Before entering the lunar lander, Shepard caught the world—and Houston—by surprise. He produced three golf balls from his pressure suit and, using the handle of a geological tool as the driver, swung at the balls. He missed the first but sent the others soaring above the terrain.

"There it goes," he commented, "miles and miles and miles."

Actually, Shepard estimated later, the first ball went about 200 yards (183 meters) and the second 400 yards (366 meters).

Apollo 14's splashdown marked the completion of the world's 40th manned space flight, the 24th flight for the United States. It restored public confidence in Apollo-Saturn equipment.

The 12-day Apollo 15 mission was designed to increase knowledge of the Moon's history and composition, plus the evolution and dynamic interaction of the Sun-Earth system.

NASA selected David Scott to command the expedition. Scott was command module pilot on the Apollo 9 mission in March 1969. His crew included Command Module Pilot Alfred M. Worden, and Lunar Module Pilot James B. Irwin.

A unique Moon vehicle, officially designated the lunar roving vehicle, but informally called the "rover," was to be used for the first time on this mission. It was fabricated by Boeing under contract to the Marshall Space Flight Center.

The rover was 10 feet, 2 inches (310 centimeters) long. Two 36-volt batteries powered tiny motors which drove each wheel. The vehicle folded into a compact unit mounted on the side of the lunar module for the trip to the Moon.

The crew set up residence at KSC in March 1971. Checkout and training for the rover had to be fitted into the schedule for the first time. A trainer vehicle equipped with tires in lieu of the mesh-type wheels of the Moon buggy was provided for familiarization driving. KSC built a traverse route, promptly dubbed "Rover Racetrack," adjacent to the lunar surface training site. During a demonstration for the press, a few reporters tried their hand with the rover trainer, guiding the vehicle through the astronauts' crater-pocked

Apollo 15 astronaut James Irwin prepares to board the lunar rover on the surface of the Moon. In the background is Mount Hadley.

sandpile. Their enthusiastic response carried over into next day's newspapers.

The crew named the Apollo command ship the "Endeavor," and the lunar module "Falcon." Target area for the mission was the rim of the canyonlike Hadley Rille in the Apennine Peaks, one of the highest lunar mountain ranges.

Beginning with Apollo 15, lunar landing missions became longer and more complex. Modifications to the spacecraft and crew support systems allowed the astronauts to double their stay time on the lunar surface. The weight devoted to lunar surface experiments also doubled. The new requirements placed a burden on KSC spacecraft and rocket engineers, especially in the area of scheduling. As one engineer commented, "... every time we powered up the ship for a major test, somebody would come down with a special requirement for their instrument."

The Apollo service module, which supplied propulsion, electrical power and environmental support for the crew, became a scientific platform as well. KSC spacecraft engineers installed experiments in a new scientific instrument bay added to the module. The gear included several kinds of cameras, a laser altimeter, spectrometers and a small satellite which would be ejected in lunar orbit. The latter could relay information on the Earth's magnetosphere and its interaction with the Moon, the solar wind, and the lunar gravity fields. These experiments were controlled by Worden as he flew Endeavor while his crewmates explored the lunar surface.

Apollo 15 rolled out to the pad on May 11. In the weeks that followed, launch crews faced an old nemesis—lightning. During the flight readiness test in June, lightning struck

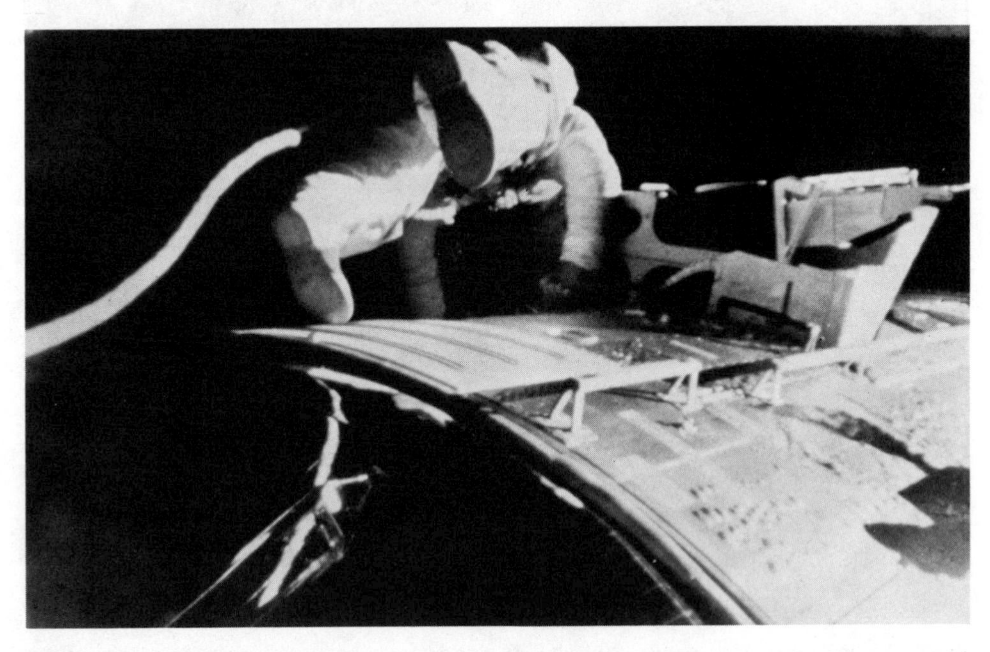

Apollo 15 Command Module Pilot Alfred Worden performed the first deep-space extravehicular activity during the return trip to Earth. Millions watched on live television.

the mobile service structure and mobile launcher, knocking out some ground support equipment. Schedules had to be revised to permit retesting of all spacecraft systems. Ten days later lightning struck again, with the same results. Damaged electrical components were replaced and spacecraft systems checked once more. During spacecraft propellant loading on July 2, a third strike caused some tests to be repeated. Lightning continued to play havoc with ground crews during final countdown, forcing Kapryan to delay moving the protective service structure from the pad until the evening before launch.

Apollo 15 lifted off precisely at 9:34 a.m. EDT, July 26. Following a routine trip to the Moon, Falcon began powered descent to the lunar surface at 6:04 p.m. EDT, July 30. Twelve minutes and 20 seconds later, Commander Scott reported, "Okay, Houston, Falcon is on the plain at Hadley."

During their lunar stay, the astronauts spent almost as much cumulative time on the Moon's surface as had all six astronauts who preceded them. The rover's performance delighted both the crew and its designers. This new mode of transportation greatly increased the astronaut's area of reconnaissance. Before liftoff, Scott positioned the rover so that its television camera could view the ascent from the surface. As Falcon's ascent stage rose, the familiar strains of the Air Force song were heard over the communications system.

Splashdown occurred Aug. 7 at 4:45 p.m., north of the island of Oahu. The impact was greater than normal because one of the three 84-foot (26-meter) diameter main parachutes failed to open properly; however, the crew was not injured. Unlike previous crews, the Apollo 15 crew was not required to spend time in quarantine.

For Apollo 16, the eighth journey to the Moon, NASA selected the Descartes region as the landing site. The choice offered the first opportunity to explore the lunar highlands, which cover about three-fourths of the lunar surface.

Navy Capt. John Young, who had flown three times in space, was selected as the commander for this mission. Thomas Mattingly, command module pilot, had been scheduled to fly on the Apollo 13 mission, but had been replaced because of exposure to measles. The final crew member was Charles Duke, the lunar module pilot.

KSC's launch team began preparing the Saturn V stages in September 1971, while the lunar, command and service modules underwent combined systems testing. The rover was installed on the descent stage of "Orion," as the crew dubbed the lunar module, in November. The command/service module was called "Casper."

Rollout occurred Dec. 13, and preparations continued for a March 17, 1972, launch. Then problems cropped up in rapid succession, making the date less certain.

Young and Duke found that their pressure suits did not allow sufficient freedom of movement, requiring modifications and retesting throughout January.

Then, factory technicians in Downey, Calif., discovered that an explosive device used to separate the command/service modules would malfunction under certain conditions. The problem could be corrected, but full-scale testing would be needed to verify the solution. A second and more serious problem cropped up immediately afterward. A fuel tank bladder in Casper's service module was inadvertently overpressurized and had to be replaced. On Jan. 27, Apollo 16 was returned to the Vehicle Assembly Building for the necessary repairs and modifications. It was the first time a flight-ready Saturn V had been removed from the pad. NASA announced a new launch date of April 16.

Working overtime and weekends, the KSC team completed the repairs in less than two weeks. Apollo 16 was returned to the pad Feb. 9. Prime backup crews took part in a simulated flight test Feb. 25, when a subsatellite which would be launched in lunar orbit was installed.

Launch countdown began at 8:30 a.m., April 10. At T minus 5 hours, 51 minutes, a gyroscope in the Saturn's instrument unit shifted abnormally for two seconds. Kapryan requested a detailed analysis. The result indicated that a shift occurring in flight would not adversely affect the mission. Countdown proceeded to liftoff at 12:54 p.m. EST, April 16.

Astronauts Eugene Cernan, commander for the pending Apollo 17, and Dr. Harrison Schmitt, the first scientist-astronaut assigned to an Apollo mission, train at KSC.

Spectators gather outside the Vehicle Assembly Building as Apollo 17, ready for the sixth U.S. lunar landing, begins its slow-motion journey to Launch Pad A.

Casper and Orion swung into orbit around the Moon a little more than 74 hours after launch. The lunar module Orion touched down in the Moon's Descartes region 860 feet (262 meters) northwest of the planned destination. Orion rejoined Casper in lunar orbit on April 23.

The crew splashed down in the Pacific at 2:45 p.m. EST, only 0.3 nautical miles (0.6 kilometers) from the planned site.

Apollo began, and Apollo ended.

The sought-after goal of a manned lunar landing had been reached. Apollo 17 was the last mission of the program and the last visit of men to the Moon for an unknown number of years. Understandably, pride was mixed with regret among the employees of aerospace companies and NASA's manned space flight organization.

The landing site for the final mission was Taurus-Littrow, named for the Taurus Mountains and Littrow Crater situated in a mountainous region of the Moon, southeast of the Serenitatis Basin. Three massifs, or rounded hills, surrounded the relatively flat target area. It was one of the most difficult landing approaches of the entire program.

Capt. Eugene Cernan, U.S. Navy, was selected to command the mission. He had flown on Gemini 9 in 1966, and was lunar module pilot on Apollo 10. Geologist Dr. Harrison Schmitt, to be his companion on the Moon, was the first NASA scientist-astronaut assigned to an Apollo mission. Navy Cmdr. Ronald E. Evans, who would also make his debut in space, was chosen as command module pilot.

At sunrise on Aug. 28, 1972, Apollo 17 atop its Saturn V was moved to Pad A.

Launch operations followed the routine established in earlier missions. Minor hardware changes, tailored to the mission's requirements, went smoothly. During pad tests, one scare arose which threatened to postpone the launch for a month. An oxidizer tank in the command module was accidently overpressurized and it was feared the tank's bladder had been ruptured. Tests showed the bladder intact, and the mission stayed on schedule for an early December launch.

Morale at the spaceport remained generally high despite budget cuts in the space program and a severe reduction in personnel—over 50 percent in Apollo launch operations. For most companies, KSC contracts would continue through the Skylab and Apollo-Soyuz missions. For the 600 members of the Grumman team, however, Apollo 17 would be the last flight for their stepchild, the lunar module. Throughout the lunar landing program, the team had gained an excellent reputation among all personnel at the center. The astronauts lauded their efforts, especially the crewmen of Apollo 13.

The Grumman crew, anxious to assure the Apollo 17 astronauts of their continued support and dedication, posted a large sign at the lunar module working level at the pad. It read: "This may be our last but it will be our best."

The last Apollo lunar flight was the first to be launched at night. Observers from hundreds of miles away reported seeing the brilliant plume of flame.

NASA Administrator James Fletcher said the slogan "should be the watchword for the entire Apollo team." It was. Morale never became a significant problem—a tribute to effective civil servant and contractor leadership and to the personal pride of the launch team members.

The launch of Apollo 17 was scheduled for 9:53 p.m., Dec. 6, the first nighttime Apollo launch. At T minus 30 seconds, the countdown suddenly stopped. Kapryan explained it to the press later: "At 2 minutes, 47 seconds before ignition, the automatic sequencer failed to provide the command to pressurize the S-IVB (third stage) liquid oxygen tank. Firing room monitors immediately observed this and took steps to perform the pressurization by manual command. At the time cutoff occurred, at 30 seconds, the stage was up to pressure and everything was normal.

"The problem was that since the terminal sequencer did not give the command, the electronic logic circuitry said that it had not happened, and so the computer stopped the count. It did not take very long to determine that we should bypass the command system, pressurize the tank manually, and continue to ignition . . . But we also had to get assurance that all functions which must occur in the last 30 seconds would indeed work properly. We wanted to avoid, at all costs, a cutoff after igniting the five engines of the first stage."

At the Marshall Space Flight Center, engineers ran through the latter phase of the countdown several times, and determined that the bypass would work.

The countdown was finally resumed, and launch occurred at 12:33 a.m., Dec. 7, 1972. The launch could be seen from hundreds of miles away. Observers in Miami, northern Florida, Georgia, and points even farther away reported seeing the plumes of flame as Apollo 17 roared into space.

Apollo 17 entered lunar orbit on Dec. 10. The next day Evans guided "America," as the command/service module was called, to a low point in orbit. Cernan and Schmitt backed away the lunar module, "Challenger." Challenger touched down on the Moon at 1:55 p.m., Dec. 11.

The astronauts explored the lunar surface during three separate excursions, collecting samples and deploying experiments. At the end of their lunar sojourn, the crew uncovered a plaque on the leg of the descent stage of the lunar module that would remain on the Moon. The plaque showed a picture of the world with a view of the Moon between the two hemispheres. It stated: "Here Man completed his first exploration of the Moon, December 1972. May the spirit of peace in which we came be reflected in the lives of all mankind." The plaque was signed by the astronauts and President Nixon.

Challenger lifted off at 5:55 p.m. EST, Dec. 14, and rendezvoused with America. Before leaving lunar orbit, the crew turned on the television camera so that Mission Control could see the dark side of the Moon. Splashdown in the Pacific occurred on schedule at 2:24 p.m., Dec. 19.

* * * * * *

What did Apollo accomplish? Scientists consider it an exciting and highly profitable exploration program. So much data was returned by the astronauts and the experiments they planted on the lunar surface that many years of analysis, evaluation and study will be required to fully digest this priceless legacy.

By studying the Moon, we have learned how to go about the business of exploring other planets. Apollo proved that we could apply to another world the methods used to understand our own. Equally important, Apollo brought about better ways of studying Earth. Because of the rarity of lunar samples, techniques were developed to measure extremely small samples, weighing only 0.00003 ounce (0.00085 gram). Also, we can now measure the ages of terrestrial rocks more accurately than before Apollo.

From geology to medicine to architecture, the extraordinary demands of space programs—Apollo in particular—spurred innovative efforts that reached into virtually every scientific and technological discipline. How this vast storehouse of new knowledge is used for the benefit of mankind is covered in Chapter 17 of this book.

What matters most about what we have learned from the Moon, in the view of many, is what it tells us about Earth—the "spacecraft" that carries over four billion of us—and the Sun, that great energy source so essential to our survival. From lunar clues as large as mammoth craters and as small as tracks etched by atomic particles into tiny crystals of soil, science is beginning to piece together the history of the solar system.

That understanding, as Apollo 17 astronaut Harrison Schmitt commented, is needed "so that we can start to tackle this long-term—50-, 100-, 200-year problem—the problem of preserving and protecting the environment of Earth."

America, the Apollo 17 command module, splashed down in the Pacific Ocean on Dec. 19, 1972.

chapter 10

SKYLAB

Skylab was the first manned effort specifically designed to develop applied space technology for improving life on Earth. Although some Skylab instruments functioned automatically, most were operated by astronauts.

The Skylab Program had several objectives:

- To enrich scientific knowledge of the Earth, the Sun, the stars, and cosmic space.

- To study the effects of weightlessness on living organisms, including the astronauts.

- To develop new and valuable processing and manufacturing techniques in zero gravity.

- To devise and test new methods of gathering information about the Earth's surface.

Comparable in volume to a modest three-bedroom home, Skylab carried at launch 2,100 pounds (953 kilograms) of food; 6,000 pounds (2722 kilograms) of water; and nitrogen, oxygen, and other life-sustaining essentials. The three three-man crews, who occupied the station for 28, 59, and 84 days, respectively, used 54 items of experimental hardware to conduct 270 scientific and engineering investigations, many of several months duration.

Skylab orbited Earth once every 93 minutes, at an altitude of 268 miles (431 kilometers). The orbital trajectory swept an area covering 75 percent of the Earth's surface, 80 percent of its food-producing regions, and 90 percent of its population. Almost 40,000 pictures were taken by Skylab's cameras.

The Skylab program began early in the 1960s, when some NASA scientists were considering how vehicles and spacecraft might be used in other projects.

On Aug. 6, 1965, NASA established the Apollo Applications Office to formulate a program that included long-duration orbital missions, during which astronauts would conduct scientific and technical experiments. The program name was changed to Skylab on Feb. 24, 1970.

As the program matured, the major elements of the station became:

- The orbital workshop, which housed the crew, most of the expendable supplies, a major experiments area, and cold gas storage and thrusters for the attitude control system. It also provided structural support for the larger solar array which would derive energy from sunlight.

- A multiple docking adapter, containing docking ports for a modified Apollo command/service module (which would transport the crew to and from Skylab), the control panel for the telescope, a window for filming Earth's surface features, and other experiments.

- An Apollo telescope mount, containing the solar experiments, control moment gyros, and its own separate solar array.

- An airlock module for extravehicular experiments, communications and data transmission equipment, the environmental-thermal system, and electrical power controls.

Marshall Space Flight Center directed the major contractual efforts, engaging McDonnell Douglas to build the workshop and airlock module, and Martin Marietta the multiple docking adapter. The telescope, photographic equipment, and food were procured by Johnson Space Center.

An Apollo-Saturn IB spacecraft, scheduled to take astronauts to Skylab, is moved to Launch Pad B, atop the pedestal adapting it to the Saturn V mobile launcher.

Two mobile launchers were modified for the Skylab vehicles. A steel pedestal 127 feet (39 meters) tall was placed on one launcher to adapt it to the Saturn IB, which was much shorter than the Apollo-Saturn Vs previously boosted from the same launcher. The second one was adapted to fit the somewhat shorter Skylab atop the Saturn V second stage.

Stages of the Skylab 1 and Skylab 2 launch vehicles began arriving at KSC during late summer

1972. For the initial launch, the Saturn V would consist of two stages plus the instrument unit. The Skylab-orbital workshop replaced the third stage.

The crew was made up of Navy Capt. Charles Conrad, the commander, a veteran of two Gemini flights and the Apollo 12 lunar landing mission; Navy Cmdr. Joseph Kerwin, a physician, and Navy Cmdr. Paul Weitz, both entering space for the first time.

The complete Skylab space station assembly was 118.5 feet (36.1 meters) long, 27 feet (8.2 meters) in diameter at the widest point, and weighed 100 tons (91 metric tons). A payload shroud provided cover for the telescope mount, airlock, and docking adapter during the boost phase.

The workshop itself contained 10,426 cubic feet (295 cubic meters) of space and weighed 78,000 pounds (35,380 kilograms). Except for attitude control thrusters, there was no propulsion system. Within the tank, aluminum open-grid floors and ceilings were installed to divide the two-story space cabin. Crew quarters in the aft end were divided by solid partitions into sleep, waste management, and experiments compartments. The thermal control and ventilation system was designed to maintain temperatures ranging from 60 to 90 degrees Fahrenheit (16 to 32 degrees Celsius), while the nitrogen-oxygen atmosphere was constant at five pounds per square inch (0.352 kg/cm2) pressure. Outside the workshop shell was a meteoroid shield to provide thermal control and decrease the possibility of hazardous punctures. Once in orbit, the shield was to be deployed and held five inches (12.2 centimeters) from the workshop shell.

Checked out and ready for mating with its booster, the orbital workshop would weigh 78,000 pounds (35,380 kilograms) and contain 10,426 cubic feet (295 cubic meters) of space.

The astronauts flew their jets into Patrick Air Force Base on May 12 to undergo final medical checks, then stood by awaiting launch. The plan was to place Skylab in Earth orbit, confirm that its systems were operating normally, then launch the first crew a day later for rendezvous and docking with the workshop.

The countdown proceeded without interruption, and liftoff occurred on time at 1:30 p.m. EDT on May 14.

Skylab's countdown went without interruption, liftoff of the Saturn V was on time, and all appeared to be going well with the launch--at first.

But, after Earth orbit was achieved, there was no confirmation that the solar arrays of the workshop had deployed. Radio signals from Earth, dispatched two hours later as a backup attempt, failed to move them. By that time the telemetry data had been examined, revealing an unexplained malfunction had occurred 63 seconds after liftoff as the vehicle reached maximum dynamic pressure. As a result, the arrays had not opened. Also, the meteoroid shield had been torn away. NASA's manned space flight managers, evaluating the alternatives, decided to delay the Skylab 2 launch until May 20, to allow time to appraise the situation.

The scope of the problem became more clear over the next several days:

- Electrical power generating capacity had been cut in half.

- Skin temperatures of the Sun side of the workshop ranged up to 295 degrees Fahrenheit (146 degrees Celsius).

- Internal temperatures hovered between 70 degrees Fahrenheit (21 degrees Celsius) on the cool side and 120 degrees Fahrenheit (49 degrees Celsius) on the wall towards the Sun.

Johnson flight controllers, in consultation with Marshall engineers, struggled to reduce the heat by changing Skylab's attitude. This required delicate maneuvering. When the workshop was tilted to minimize solar exposure, the electrical supply dropped. But if the solar wings were tilted toward the Sun, the workshop became hotter.

By May 17, Skylab was stabilized at an angle of about 50 degrees to the Sun. Temperatures in the workshop fell to about 90 degrees Fahrenheit (32 degrees Celsius). Astronauts Conrad and Kerwin of the Skylab 2 prime crew were at Marshall, rehearsing space walk procedures to install a sunscreen to replace the missing meteoroid shield.

The manned launch was further delayed until May 25, to allow time to develop a shield and the means to deploy it. Additional cryogenics were loaded into the service module tanks, to extend the normal life of the fuel cells so that they could supplement Skylab's electrical budget.

Within hours of launch, Johnson delivered a parasol device fashioned from aluminized Mylar-nylon laminate. It could be deployed through the scientific airlock like an umbrella, opened up in space, and pulled down into position.

The countdown clock for Skylab 2 had been stopped at T minus 59 hours. It was resumed at 8:30 p.m. EDT, May 22. The Saturn IB roared off from Pad B at 9 a.m. EDT, May 25, and Conrad steered his ship to the rendezvous point in space. Late in the afternoon the crew approached Skylab and turned on television equipment so engineers on Earth could visually assess the situation. Most of the micrometeoroid shield had disappeared, but some remnants were jammed against the remaining solar wing, preventing its deployment. The other wing had completely vanished.

Conrad guided the ship to a soft dock with the laboratory, and the crew attempted unsuccessfully to free the solar wing.

After undocking, the Skylab 2 crew photographed their repair job. The solar shield they installed during extravehicular activity is visible on the orbital workshop.

They abandoned the attempt, then sought to hard dock with Skylab, only to encounter a problem with the latching mechanism. After several unsuccessful attempts, they removed the drogue—working outside the command module—and trimmed up the latch system, finally locking the spacecraft firmly with Skylab.

Entering the workshop next day they found the interior hot but otherwise shipshape and proceeded to install the parasol device. Within hours the temperature began to drop; by May 29 it had reached 83 degrees Fahrenheit (28 degrees Celsius). The crew turned on experiments, began recording solar phenomena with the telescope, and filming Earth as planned. Subsequently, Conrad and Kerwin performed another work-in-space task—and this time freed the jammed solar wing.

The crew completed the planned 28 days in space and accomplished the objectives required by 48 experiments. The astronauts also filmed a hurricane in the Pacific Ocean, and a solar flare. They disengaged from their temporary home in space and landed in the Pacific, southwest of California.

Having carefully reviewed Skylab's status following the mission, Program Director William C. Schneider announced that the second mission was planned to extend for

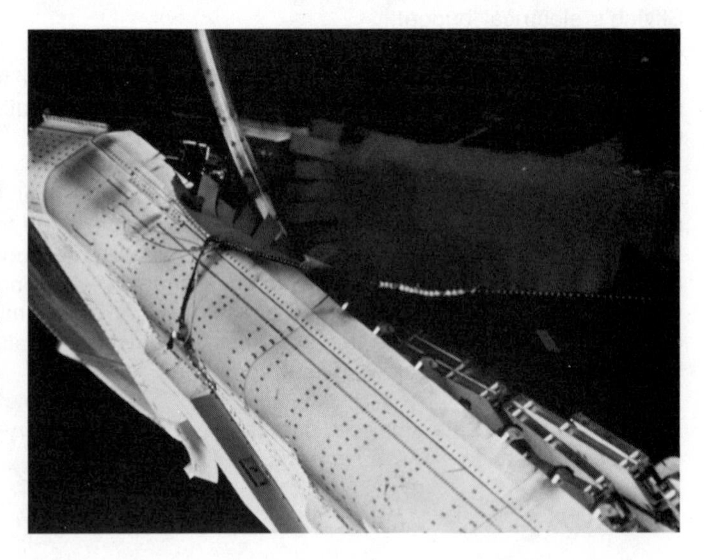

A close view of the solar panel shows the aluminum strap that prevented deployment. The opposite panel was torn from its mount when the meteoroid shield deployed prematurely.

56 days, thus doubling the astronauts' exposure time. NASA's physicians would check the crew's health daily. At the end of 28 days, the doctors would thereafter decide on a weekly basis if the crew could continue.

Navy Capt. Alan Bean, veteran of Apollo 12, commanded the Skylab crew. The science pilot was Dr. Owen Garriott, a specialist in ionospheric physics, having been trained by NASA as a jet pilot. Maj. Jack Lousma, U.S. Marine Corps, completed the crew.

Preparations for the mission were in progress at KSC when NASA decided to attempt a full-duration mission.

Skylab's problems prompted Schneider to announce on June 6 that the Skylab 3 launch would be advanced from Aug. 8 to July 28 to get the astronauts into the space station at the earliest possible time. Accordingly, KSC Launch Director Walter Kapryan tightened the work schedules.

Numerous thundershowers occurred the afternoon of July 27. Extensive fog reduced visibility to three miles (4.8 kilometers)—the distance between the pad and the Launch Control Center, with ground fog at 600 feet (183 meters) in spots. Still, the terminal countdown began at 7 a.m. EDT on July 25, and liftoff came at 7:11 a.m. EDT, July 28.

Skylab 3 vanished in the fog shortly after launch. Five hours later the crew steered to a firm docking with Skylab and moved in to occupy the workshop, In retrospect, NASA physicians thought the relatively quick transfer caused the motion sickness which disabled all three astronauts soon after they took over Skylab. The Skylab 2 crew possibly escaped the problem because they slept in the command module the first night and carried out a more gradual transition into the space station.

By the fourth day, the astronauts' recovery from the motion sickness was complete, and they started up the experiments and carried out the planned routines. On Aug. 6, Garriott and Lousma installed a new sail, replacing the parasol. The internal temperature was soon stabilized.

For their second task performed outside the station, Lousma installed and connected six rate gyroscopes carried aloft in the Skylab 3 spacecraft. The gyros replaced those which had malfunctioned in the months following Skylab's launch, possibly because they were baked by temporary high temperatures after the loss of the meteoroid shield.

From time to time during the mission, television viewers on Earth saw the astronauts go about their duties. From the 29th day on, new records for time-in-space were established daily.

Garriott took special delight in operating the telescopes which filmed solar phenomena, since the Sun continued very active in a period of supposed quiet. More than 100 solar flares were observed. Dr. Neil R. Sheeley of the Naval Research Laboratory reported that Garriott witnessed magnetic field phenomena, never seen by observers using Earth telescopes.

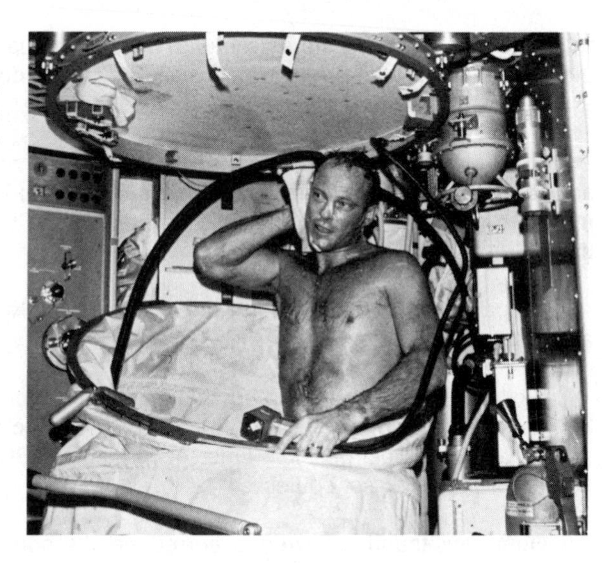

Skylab 3 Pilot Jack Lousma enjoys one of the physical comforts absent from earlier, more cramped space missions--a hot bath.

After tidying up their space home at the end of their visit, the crew undocked on Sept. 25 and splashed down in the Pacific Ocean at 6:19 p.m. EDT, in rough seas.

Skylab experimenters were enthusiastic about the quality of Skylab data, and were busy evaluating the information in terms of its potential applications.

Skylab's third crew was unusual in that all three astronauts were making their first trip into space. Gerald Carr, the commander, was a Marine Corps lieutenant colonel. Dr. Edward Gibson, science pilot, had joined NASA in 1965. The agency gave him pilot training. Lt. Col. William Pogue, U.S. Air Force, rounded out the crew in the pilot's berth.

A leaking coolant loop in Skylab was causing concern. Marshall engineers foresaw the possibility of the system running out of coolant by about Nov. 20, unless repair and replenishment fixed the problem. So there was some urgency attached to the prospective launch date of Nov. 10. A repair kit would be carried aboard Skylab 4.

The flight readiness review considered a matter of new concern, stress corrosion in the

Astronaut-scientist Owen Garriott is pictured as he deploys an experiment during an extravehicular activity at Skylab's Apollo telescope mount.

forged aluminum "E" beams of the first stages of the Skylab 4 vehicle and the standby Skylab rescue vehicle. The beams were load-bearing and supported the structure. The cracks formed by fatigue, corrosion, and load carrying were repaired by removing metal on either side and welding new metal in the opening.

Stress corrosion popped up again Nov. 6. This time hairline cracks appeared in eight large tail fins of the first stage. Fourteen cracks were discovered. The fins helped stabilize the rocket in flight and also supported its weight on the pad.

The corrosion may have resulted from increased salt concentration in the air at the oceanfront launch pad. Kapryan recommended delaying the launch until Nov. 16, allowing time to bring in spare fins from Michoud, La. Contractor crews worked around the clock applying doublers on the new fins to strengthen their attachment to the stage.

Two days before launch, shallow, almost indistinct cracks were found in aluminum beams within the second stage. Experts in metallurgy and fracture mechanics concluded that the cracks would not adversely affect the beams. The Manned Flight Management Council accepted the findings that the stage was safe for flight.

The Management Council also approved the proposal to consider Skylab 4 a 59-day mission, which could be extended week by week to a maximum of 84 days. This involved loading more food in the Apollo spacecraft, including an enriched foodstuff to stretch out the primary food supply. Since the Skylab 4 crew would exercise more, physicians concluded there would be no lasting ill effects from spending virtually three months in zero gravity.

On Nov. 16, 1973, Skylab 4 lifted off the mobile launcher at Pad B at 9:01 a.m. EST. The Saturn IB performed flawlessly, and rendezvous with Skylab was accomplished on the fifth revolution. The crew encountered some difficulty in docking but made the hard connection on the third attempt.

As soon as they adjusted to the Skylab environment, the crew went to work, repairing and servicing the coolant systems of the airlock module. They settled into a full routine of tending experiments, exercising, eating and discussing schedules and notable sightings with Mission Control.

By early December, the crew's productive routine included photographing solar phenomena and comet Kohoutek, and measuring Earth resources. Infrared cameras and heat-seeking sensors probed the Earth from 275 miles (443 kilometers) in space, looking for evidence of oil deposits and seeking natural steam wells which might be harnessed to yield power in areas of California, Arizona, New Mexico, Central America, and Mexico.

The astronauts completed their first month in space on Dec. 14, surpassing the time of the stay of the Skylab 2 crew.

Christmas Day was a busy time for the astronauts, who spent seven hours photographing comet Kohoutek and performing other experiments outside the workshop.

During early January, they continued to observe Kohoutek, conducted Earth resources studies, filmed solar phenomena and managed a wide range of other experiments.

On Jan. 14 the three astronauts broke the Skylab 3 record of 59 days, 11 hours, and 9 minutes in space. On Jan. 25 all three crewmen broke Alan Bean's record of 69 days in space—exchanging congratulations with Bean, who sat in Mission Control. Program Director Schneider warned that an ailing gyroscope might cease to function at any time. However, it might be possible to complete the mission by using the gas thrusters and the Apollo thruster system to stabilize Skylab. As a precaution, NASA ordered the USS New Orleans, the recovery ship, into the recovery area three days early.

Near the end of their stay, the astronauts packed gear in Apollo and put the space station in order for a possible visit by astronauts in the future. On Feb. 8, they undocked the Apollo spacecraft, said their goodbyes to Skylab, and headed for splashdown. They had traveled some 34 million miles in their space station. The recovery went without a hitch.

Carr, Pogue, and Gibson met the press on Feb. 22, showed films which sampled their voluminous coverage of Earth terrain features, and talked of the highlights of observations of the Sun and comet Kohoutek.

There were many honors for the Skylab crews and for the men and women who designed and built Skylab, who prepared the vehicles and spacecraft for launch, and who managed the mission so effectively.

The Skylab space station, many of its systems still operating, remained in orbit for another 5 1/2 years. At 12:39 p.m. EDT, July 11, 1979, it re-entered Earth's atmosphere and disintegrated, raining debris harmlessly along a path extending over the southern Indian Ocean and sparsely populated areas of western Australia. Concern over the huge spacecraft's impact point was one of the most talked-about events of the young space age, but in the end, it proved groundless. The Skylab mission began with major problems and ended with much speculation about its return to Earth, but in fact the program established an unparalleled record for scientific return on the investment. Skylab paved the way for the Spacelab, the major scientific payload of the Space Shuttle.

chapter 11

APOLLO-SOYUZ

Even to space buffs who had avidly followed telecasts of the Apollo and Skylab astronauts in space, the broadcasts from Earth orbit on July 17 and 18, 1975, were unusual. First, they saw three U.S. astronauts in an Apollo spacecraft rendezvous and dock with a Russian Soyuz spacecraft manned by two cosmonauts. Then during two days of mated flight, they observed the crew members move from one craft to the other, share meals, exchange gifts and conduct scientific experiments.

The event was the Apollo-Soyuz Test Project—history's first international manned space flight. The basic purpose of the nine-day joint mission—to flight-test a mechanism for joining two orbiting spacecraft—was successfully met. Apollo-Soyuz demonstrated that international space rescue missions would be feasible in the future.

Scientists also took full advantage of the opportunity for more orbital research projects, conducted by each crew independently. Apollo carried equipment for 23 science and technology experiments; Soyuz was equipped for six.

The two spacecraft were outfitted for five joint experiments as well, three carried out while the two spacecraft were linked together and two after Apollo and Soyuz had undocked.

Although Apollo-Soyuz was brief compared to the earlier Skylab flights, negotiations and preparations stretched over several years. As far back as the late 1950s, shortly after the launchings of the first artificial Earth satellites, both the Soviet Union and the United States endorsed the principle of international cooperation in space.

In October 1970, representatives of NASA and the Soviet Academy of Sciences met in Moscow for the first of a series of sessions, later held alternately in the Soviet Union and the United States. The conferences determined that a joint space venture might take the form of a rendezvous and docking in Earth orbit.

On May 24, 1972, President Nixon and Alexei Kosygin, chairman of the U.S.S.R. Council of Ministers, signed an agreement in Moscow concerning "cooperation in the exploration and use of outer space for peaceful purposes." The agreement approved the idea for an Apollo-Soyuz test flight.

Preparation began on both sides. Mission managers and astronaut crews were selected. As in previous Apollo and Skylab missions, the U.S. crew consisted of three astronauts; two cosmonauts formed the U.S.S.R. crew. Training began almost immediately.

The American prime, backup, and support crews—nine men in all—visited Star City near Moscow to train with the Soviets. Soviet crews visited the Johnson Space Center for training and flight simulation sessions and toured the launch facilities at KSC. While in Florida they went to Disney World with the U.S. astronauts and rode on one of the main attractions, appropriately called "Space Mountain."

Technical experts from both nations held 44 meetings in the two countries over a three-year period. Scale models and exact duplicates of the different docking mechanisms built by each country were docked and undocked hundreds of times under simulated space conditions.

To accommodate the docking systems and to provide a chamber through which crew members could pass from the atmosphere of one craft to the different atmosphere of the other, an airlock, called the docking module, was designed and built by the United States. Attached to Apollo during most of the mission, it was jettisoned before the spacecraft re-entered Earth's atmosphere. Basically, the docking module was a cylindrical aluminum corridor 10 feet, four inches (3.15 meters) long and four feet, eight inches (1.4 meters) wide at its greatest diameter.

Soyuz had a two-gas atmosphere—nitrogen and oxygen—about the same as the atmosphere of Earth at sea level, with a pressure of 14.7 pounds per square inch (1 kilogram per square centimeter). Apollo's atmosphere was pure oxygen with a pressure of 5 pounds per square inch (0.35 kilograms per square centimeter). Before docking, plans called for the Soviets to reduce the atmospheric pressure inside Soyuz to 10 pounds per square inch (0.7 kilograms per square centimeter) in order to shorten the time the crew needed to remain inside the docking module to safely transfer from Soyuz to Apollo, from about two hours to a few minutes.

One major worry was the language barrier. Obviously, successful operations and the safety of the crews demanded speedy, accurate exchanges of information—both in the control centers and in orbit. As part of their training, astronauts and cosmonauts studied each other's language. During the joint phase of the flight, the U.S. crew spoke in Russian and the Soviet crew in English.

A dictionary of common terminology was compiled; translators largely memorized it. Also, a large volume containing joint operating instructions was published and placed aboard each spacecraft. Half of each page was in English; the other half, with the same text, was in Russian.

A unique communications feature for the upcoming mission was the utilization of NASA's ATS-6 (Applications Technology Satellite). Without the satellite, because of their relatively low orbits, the two spacecraft would have been in line-of-sight contact with Earth stations for only about 17 percent of the time. However, with use of the ATS-6 in geostationary orbit, communications between Earth and the manned spacecraft were possible for about half of each 88-minute orbit, and millions of people in many parts of the world were able to see and hear much of the mission.

While the crews and various support personnel pushed ahead with their training for the mission, launch preparations at KSC accelerated during the latter part of 1974 and continued until final launch countdown began on July 11, 1975.

The Saturn IB vehicle had been employed for Earth-orbital Apollo test flights prior to the Moon launches, and served as the launch vehicle for the three manned Skylab missions.

The first stage was removed from its storage "cocoon" in the Vehicle Assembly Building on Nov. 27, and stacked on its mobile launcher in mid-January. The following day the second stage was placed atop the first stage; two days later the instrument unit was added.

Major modifications to the Apollo command/service module for Apollo-Soyuz were extra propellant tanks for the reaction control system, additional equipment to operate the docking module and its docking system, and provisions for experiments. After its arrival, the command/service module underwent testing in the Operations and Checkout Building until early in 1975.

The docking module was mated to its docking system, and the complete unit placed in the spacecraft adapter in mid-February. The adapter originally had been used on Moon missions to house the lunar module until shortly after translunar injection began; for manned Skylab flights, the adapter had been essentially empty.

Mating of the command/service module and the spacecraft adapter took place in late February and early March. On March 9, the launch vehicle was crowned with this assembly. Then the completely assembled Apollo-Saturn IB was moved to Pad B at Launch Complex 39 on March 24.

After the vehicle was mounted atop pedestals at the pad, final preparations began, culminating with the countdown demonstration test which started on June 25 and ended July 3. As scheduled, the final launch countdown started on July 11.

Modifications to the launch pad were minimal. The most visible change was the installation of a new lightning protection system, a fiberglass mast atop the mobile launcher from which the vehicle would be launched. Over this mast a cable was passed to grounding points on each side of the mobile launcher. By providing a point of stroke impact, it kept the mobile launcher and the space vehicle from serving as a current path to the ground. The path provided was separated and insulated from all vital equipment.

Launch was scheduled at a time of day and season of the year with a history of thunderstorm activity (3:50 p.m. EDT, July 15). The launch window was approximately 10 minutes long; windows on the following four days were slightly less. There would be five opportunities for launch, based upon a maximum mission time of six days for the Soyuz spacecraft and a nominal liftoff of Soyuz at 8:20 a.m. EDT on July 15.

While final hardware preparations moved ahead at KSC and the Cosmodrome at Baykonur, 2,000 miles (3,219 kilometers) southeast of Moscow, the two crews readied themselves.

As one veteran reporter noted about the two crews, Apollo-Soyuz would be a triumph for middle age because all five crew members were over 40.

Apollo-Soyuz crewmen, touring KSC are, left to right, Apollo astronauts Donald "Deke" Slayton, Vance Brand and Thomas Stafford, and Soyuz cosmonauts Valeriy N. Kubasov and Alexei A. Leonov.

Thomas Stafford, 44, then a brigadier general in the U.S. Air Force, commanded Apollo. On three previous flights, he totaled over 290 hours in space. He was considered the world's leading expert in the two key maneuvers required to join Apollo and Soyuz—rendezvous and docking.

His companions aboard Apollo were Vance Brand, 44, and Donald "Deke" Slayton, neither of whom had flown in space before.

Slayton, 51, became the oldest person to fly in space at that time. Although one of the original seven astronauts selected in 1959, he was the only one of that group never to have been in space. Chosen to pilot a Mercury mission in 1962, he was replaced and disqualified for space flight status when doctors discovered an irregular movement in his heart muscles. After his heart irregularities stopped, he was restored to flight status in 1972 and subsequently assigned to the Apollo-Soyuz mission.

Pictured with their Saturn IB launch vehicle at Complex 39 are, from left to right, Apollo-Soyuz astronauts Thomas P. Stafford, mission commander; Donald K. "Deke" Slayton; and Vance D. Brand.

Commander of the two-man Soyuz crew was Soviet Air Force Col. Alexei A. Leonov, 41, a cosmonaut since 1960. In 1965 he had been the first man to "walk in space." His flight engineer was Valeriy N. Kubasov, 40, who had acted as the flight engineer on the Soyuz 6 mission in 1969. During that flight, he became the first person to weld materials in space, under conditions of weightlessness.

The Soyuz spacecraft had completed more than a dozen successful flights in space. Overall, it was smaller than Apollo, and weighed about half as much as the U.S. craft. Extending from opposite sides of Soyuz were two winglike solar panels that converted sunlight to electricity to recharge its batteries. Apollo was powered by fuel cells, eliminating the need for solar panels.

On July 15, 1975, despite the difficulties of language and distance in coordinating the two missions, each was launched within a fraction of a second of its scheduled time. Thus began the 31st manned space flight for the United States, the 27th for the U.S.S.R.

At Baykonur, this Soviet launch was the first ever shown on live television. Another first was the invitation, and attendance, of U.S. officials to view the liftoff. The U.S. ambassador to the Soviet Union, Walter J. Strossel, was among those watching.

At KSC, the Soviet ambassador to the United States, Anatoly F. Dobrynin, watched with NASA Administrator James Fletcher as Apollo lifted off.

After each spacecraft achieved orbit, a series of complex maneuvers by Apollo brought it to the vicinity of Soyuz. On July 17, the astronauts maneuvered Apollo into formation flight with Soyuz and then linked them together.

America's three members for Apollo-Soyuz begin their historic flight, proof that international rescue missions in space are possible.

The historic union took place at 12:09 p.m. EDT, above the Atlantic Ocean west of Portugal.

"Docking is complete, Houston," announced Stafford.

"Well done, Tom," came an accented voice via radio from Soyuz.

For the next two days, crew members crawled through the docking module exchanging visits, various mementos, and flight rations. Included in the Russian meal given to the astronauts were borscht, jellied turkey, and black bread. All experiments scheduled for the docked period of the mission were successfully conducted.

On July 19, almost 44 hours after they had initially docked, the two craft separated for about 30 minutes. Then a second docking was conducted to gain further experience. Three hours later they separated for the final time.

For one of two joint experiments to be conducted after docking, Apollo was backed away in a straight line until its circular outline blocked out the brilliant center of the Sun and gave the Soyuz crew a chance to photograph the solar corona. In effect, Apollo created an eclipse of the Sun for Soyuz.

Two days after the final undocking, Soyuz successfully landed in Russia. Its landing, like its launch, was shown live on television for the first time.

The Apollo descent to the Pacific three days later was more eventful. Just minutes before splashdown, the crew noted a yellow gas in the capsule that caused eye irritation and severe coughing. This was not a serious concern at the time since other Apollo crews had reported fumes and odors during re-entry. Compounding the situation, however, the command module flipped upside down upon hitting the water. This, too, had happened on other flights. But, for about five minutes, the astronauts hung upside down in their seats. Stafford finally wriggled free of his seat straps and reached the oxygen masks stored on board. Brand, meanwhile, had fainted. He gained consciousness after about a minute.

The craft was righted by flotation balloons, and the hatch opened to let fresh air in. Then the crew and the capsule were lifted to the deck of the recovery ship, the USS New Orleans. Although the crew seemed comfortable at the welcoming ceremonies, they complained of chest pains during the postflight physical.

Investigators later determined that the crew had failed to activate the Apollo's automatic Earth Landing System. As Apollo bored into the atmosphere, the crew manually activated its parachutes, but failed to turn off the small rocket thrusters that position the capsule before the parachutes open. Thruster cutoff would have occurred automatically if the Earth Landing System had been activated.

One thruster outlet was near the intake through which fresh air was sucked into the capsule during its descent and after splashdown. Thus, the thruster propellant's oxidizer was allowed to enter the spacecraft and the gas the astronauts saw, smelled and inhaled was what had been feared the most by the physicians—nitrogen tetroxide, a very toxic chemical agent that can be fatal if inhaled in large concentrations.

After 13 days of medical observation in Hawaii, each crew member was given a clean bill of health as far as exposure to toxic gas was concerned.

Concern for the health of the crew resurfaced on Aug. 19 when it was announced that Slayton would undergo surgery to explore a spot on his lung. Fortunately, the lesion was benign. Doctors said it had developed before the mission.

When Apollo splashed down, it closed a chapter in the history of America's space program. It was the last flight for Apollo—the vehicle which had carried 45 Americans through space in 15 separate flights over a span of seven years. Its successor, the Space Shuttle, can complete numerous round-trip missions between Earth and Earth orbit.

The era of the Space Shuttle finds many citizens of different nations working together in orbit. Apollo-Soyuz provided a preview of such a scenario.

chapter 12

SPACE SHUTTLE

A new concept was born of the need for a reusable vehicle. This new "spaceship" would not be a ballistic missile with people. It would have wings, and would fly home.

As Apollo moved into high gear, NASA began to look at the directions space programs of the future might take. On Sept. 15, 1969, a Space Task Group appointed by President Nixon recommended broad outlines for the next 10 years of space exploration. Among the options was the concept of a reusable Space Shuttle, which offered major advantages over conventional rocket systems. It would be reusable for up to 100 or more missions. Moveover, spacecraft built to fly aboard a Shuttle vehicle could be designed with more emphasis on their mission capabilities, and less emphasis on their ability to withstand the rigors of conventional rocket launches. In addition, studies indicated a reusable Space Shuttle would require the least capital risk per flight, offer the lowest technical risk in development and provide the highest rate of return on the government's investment. The U.S. Air Force, which had been actively involved with Shuttle planning and which would also participate in Shuttle missions, agreed with this judgment.

The design for the new Space Shuttle vehicle called for three attached main elements: an orbiter, which carries astronauts and payloads into orbit; an external propellant tank; and two unmanned solid rocket boosters. The boosters burn in unison with the orbiter's three main engines, providing the primary thrust to get the Shuttle off the ground. The propellant in the two boosters burns out about 29 miles (47 kilometers) above Earth. At that time, the booster casings separate from the orbiter and fall into the sea, their descent being slowed by parachutes. Specially equipped ships in the recovery area retrieve the floating casings and parachutes. The casings are then towed back to the launch area for refurbishment and reuse.

To save space and weight on the orbiter, the tanks for the propellants which power its engines are not part of its fuselage. Instead, the design incorporates a large external tank, divided to hold both the liquid hydrogen and the liquid oxygen—a total of more than 1,577,000 pounds (715,300 kilograms) of propellants. The tank is jettisoned over a remote part of the ocean just before the orbiter enters orbit. It is not recovered.

The 122-foot (37-meter) long orbiter, about the size of a DC-9 jetliner, continues on its Earth-orbital mission for a period of three to 10 days. It then re-enters the atmosphere, and like an enormous glider, makes an unpowered landing on a runway. Then the orbiter is refurbished, assembled with refurbished boosters and a new external tank and checked out for a new mission. Orbiters are designed to fly up to 100 missions, solid rocket casings for 20 or more missions.

The Shuttle concept represents a whole new way of space flight. On a standard mission, it carries up to seven crew members, only three of whom need be NASA astronauts. The others can be payload specialists, usually technicians, engineers or scientists, who make observations and conduct experiments. For rescue missions, the orbiter's cabin holds as many as 10 persons; this means that an orbiter with a basic three-man crew can rescue all occupants of a disabled orbiter.

The orbiter's cargo bay measures 60 feet (18 meters) by 15 feet (5 meters) and carries up to about 55,000 pounds (24,948 kilograms) of payload, but because of constraints imposed should a mission have to be aborted and the orbiter return to Earth still carrying its cargo, the maximum payload capability is capped at about 50,000 pounds (22,680 kilograms).

Multiple payloads (up to five satellites) may be carried in the cargo bay. Fully equipped laboratories, where experimenters work in a shirt-sleeve environment, can also be accommodated. Specialists can repair damaged satellites already in orbit, or, if necessary, return the satellites to Earth for a major overhaul. For missions requiring higher orbits than the orbiter's maximum altitude of about 500 miles (805 kilometers), a small solid rocket stage is attached to the satellites carried aboard. After being unloaded and checked out in space, the solid rocket is used to boost the satellite into the higher orbit. Likewise, interplanetary spacecraft or deep space probes are accelerated into their trajectories by this procedure.

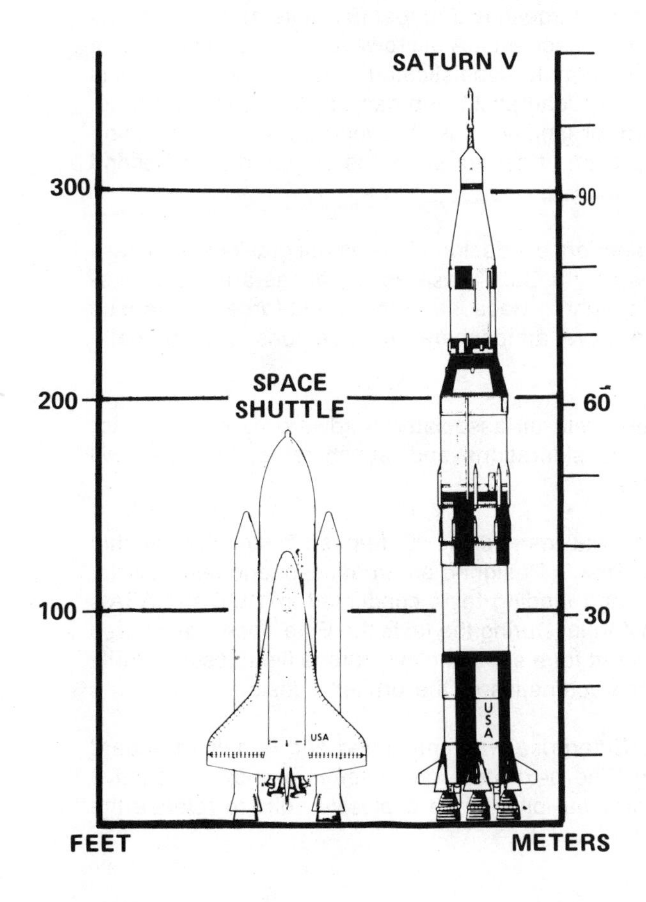

The powerful Saturn V could loft a 250,000-pound (113-metric ton) payload into near Earth orbit, but all components of the launch vehicle could only be used once. The Shuttle can take a 50,000-pound (22,680-kilogram) payload into near Earth orbit, and all components except the external fuel tank are reused.

Once it enters Earth orbit, the orbiter uses small rocket engines in its orbital maneuvering system to adjust its path, to facilitate rendezvous operations, and, at the completion of orbital operations, to slow down for re-entry into the atmosphere.

A reaction control system containing small thruster rockets provides the Shuttle orbiter with attitude control in space, and precision velocity changes for the final phases of rendezvous and docking or orbit modification. Used in conjunction with the aerodynamic control surfaces, it provides attitude control during re-entry.

The orbiter does not follow a ballistic path to the ground like previous manned spacecraft. It "pancakes" into the Earth's upper atmosphere with its nose pointed up at a steep angle. At lower altitudes, the orbiter goes into a more nearly horizontal flight for an aircraft-type approach and landing. The orbiter has a cross-range capability (can maneuver to the right or left of its planned entry path) of about 1,270 miles (2,044 kilometers). Landing speed is about 215 miles (346 kilometers) per hour.

The orbiter touches down like an airplane, on runways at either Kennedy Space Center or the Dryden Flight Research Facility in Edwards, Calif.

At KSC, the orbiter lands on a 15,000-foot (4.6-kilometer) runway. Located northwest of the Vehicle Assembly Building, the Shuttle Landing Facility has a northwest-southeast alignment. This airstrip, one of the world's largest, is 300 feet (91 meters) wide and has a 1,000-foot (305-meter) safety overrun at each end. A microwave beam landing system guides the orbiter to a landing. An extremely sophisticated and accurate system is necessary because the orbiter makes a "dead-stick" approach to the runway; that is, it has no flight power system on board for landing. It is, in effect, a glider at this point. In the unlikely event of a missed approach, it cannot circle the strip and try a second time.

During re-entry, crew members experience a designed maximum gravity load of less than 1.5 Gs; during launch it reaches only 3 Gs. (These accelerations are about one-third the levels experienced on Apollo flights.) Because of the low G force and various other features, such as a standard sea-level atmosphere, most persons in good health can safely ride aboard the orbiter.

A fleet of four operational orbiters, with all associated hardware, is projected for the Space Shuttle program. Budget considerations and launch rates, however, will determine the final number of orbiters.

The first orbiter to be constructed was named the "Enterprise," after the flagship in the popular television series, "Star Trek." Designed as an atmospheric test vehicle, it was used for the Shuttle approach and landing tests conducted in 1977 at NASA's Dryden Flight Research Facility in California. During the tests the Enterprise was carried atop a modified Boeing 747 carrier aircraft for a series of low-altitude flight tests to verify the aerodynamic and flight control characteristics of the orbiter's design.

During the initial test flights, the Enterprise was unmanned and remained aboard the 747 from takeoff through landing. The next step was a series of manned captive flights. For the final series, NASA astronaut-pilots fired explosive bolts to release the orbiter from the 747, and flew it to an unpowered landing several minutes later. All tests were successfully completed ahead of schedule in October 1977.

Following the approach and landing test program, Enterprise was shipped to Marshall in Huntsville, Ala., for ground vibration tests. The whole configuration—orbiter, external tank, and solid rocket boosters—was tested to see how it could withstand the stresses incurred at launch. Following these tests, the Enterprise was ferried to the Kennedy Space Center where it was used as a facilities verification vehicle.

The Enterprise was the first orbiter to fly--in verification, approach and landing tests. Designed as a test vehicle, it would never leave Earth's atmosphere.

The second orbiter built was named the "Columbia," after the American Naval vessel that circumnavigated the globe in the 18th century. The first orbiter scheduled for space flight, Columbia was delivered to Kennedy in March 1979, and began flight processing for its first launch, which occurred April 12, 1981 (see next chapter). By the end of 1985, three more orbiters had arrived at Kennedy: Challenger, Discovery, and Atlantis. Challenger was destroyed during a launch failure in 1986, and will be replaced by Endeavour, scheduled to arrive at KSC in 1991.

* * * * * *

As the Space Shuttle concept was being developed, NASA assigned areas of program responsibility to its Centers. KSC was given the responsibility for designing ground support facilities and systems for the Shuttle. The Johnson Space Center became lead center for the Shuttle program and was responsible for designing and procuring the orbiter. The Marshall Space Flight Center was charged with design and procurement of the external propellant tank, the three main engines of the orbiter, and the solid rocket boosters.

Just two days before the launch of Apollo 16 on April 14, 1972, Dr. George M. Low, acting NASA administrator at the time, announced that KSC would be the initial launch site for the Shuttle. The Department of Defense later was authorized to construct a second site at Vandenberg Air Force Base in California, to handle launches into polar orbits (across the North and South Poles). However, the 1986 Challenger accident and subsequent hiatus in Shuttle launches resulted in the moth-balling of the California facilities.

KSC's future for manned missions was assured. Many of the same structures originally constructed and equipped for Apollo would serve for the Shuttle in their present configuration; others required varying degrees of modification. Soon after the official announcement, KSC geared up to build the Shuttle Landing Facility and other facilities unique to the needs of Shuttle operations. Ground-breaking ceremonies for the three-mile (4.8-kilometer) runway took place in April 1974.

With a view to keeping costs down, planners took full advantage of existing structures and scheduled new construction only when a unique requirement existed. One of the first major new buildings was the Orbiter Processing Facility, located in the heart of Complex 39. It connects with the Shuttle Landing Facility and the nearby Vehicle Assembly Building by tow ways similar to aircraft taxiways.

The Orbiter Processing Facility is essentially a hangar with two high bays in which orbiters undergo "safing" and servicing after landing. It is here in a "clean-room" environment that the propellant feedlines are drained and purged and explosive actuators removed. Next, flight and landing systems are refurbished, returned payloads are removed, and the payload bay support equipment is inspected and refurbished. If the payload scheduled for the next mission is to be installed horizontally, it will be placed in the orbiter while it is in the hangar. (Other payloads of the vertical-placement type are installed in the cargo bay after the vehicle is in position on the launch pad.)

Two orbiters in parallel flow can be handled in the Orbiter Processing Facility. In 1987, a third orbiter processing facility was opened nearby. The Orbiter Modification and Refurbishment Facility has a high bay of the same dimensions as the two in the Orbiter Processing Facility, but at present only non-hazardous work can be performed here. NASA plans to upgrade the Orbiter Modification and Refurbishment Facility to allow total orbiter checkout like that conducted in its sister hangars.

Because Shuttle vehicles differ significantly in size and shape from previous manned space vehicles, a technological "face-lift" was undertaken to prepare some existing structures for their new roles. Modifications in the Vehicle Assembly Building included major changes to High Bays 1 and 3 to equip them for the assembly and checkout of complete Space Shuttles. Work platforms had to be reshaped to fit the Shuttle configuration.

The other two High Bays, 2 and 4, required internal structural changes to accommodate one vertical storage cell and one checkout cell apiece. The 154-foot (47-meter) Shuttle external tank, shipped by barge from Louisiana, is brought here.

The unique makeup of the Shuttle vehicle demanded some changes to facilities designed for the Apollo-Saturn. The external fuel tank, largest Shuttle component, needed to be accommodated. The tank then had to be mated, successively, with the solid rocket boosters, below left, built up from segments shipped to KSC, and finally with the orbiter, below right.

The Launch Control Center built for the Apollo program was upgraded to conduct Shuttle launches with the addition of the KSC-designed Launch Processing System (LPS). The LPS computer network automatically performs many countdown activities which formerly would have required additional manpower to perform and monitor.

A low bay checkout cell was converted into an enclosed, environmentally controlled workshop where orbiter main engines are received and inspected. The workshop also serves as a support facility for all main engine operations at the Center.

In addition, the north door of the huge assembly building was widened 40 feet (12.2 meters) to accommodate the 78-foot (24-meter) wingspan of the orbiter as it is towed into the building prior to assembly with other Shuttle elements. The widest stage of the Saturn V was 33 feet (10 meters) in diameter.

Two of the four firing rooms in the adjacent Launch Control Center were equipped with consoles, computers and associated equipment that are one element of the Launch Processing System. This advanced system performs much of the checkout of the Space Shuttle vehicle while the vehicle components are being processed for launch. It was especially developed at KSC for the Shuttle program. With the Launch Processing System, the final countdown for the Shuttle was compressed from the 28 hours needed for an Apollo launch to three hours.

The launch pads, too, needed alterations. Here, Pad A, undergoing water deluge tests, displays its new fixed service structure and attached rotating service structure.

The Apollo launch pads at Complex 39 also underwent major changes. All the structures on the surface of the pads were removed with the exception of the six fixed pedestals at each pad. For previous manned missions, these pedestals were used to support a mobile launcher holding an Apollo-Saturn vehicle. The same pedestals hold the modified mobile launcher platforms on which Shuttle vehicles are assembled.

The most obvious change to the mobile launchers used for previous Saturn vehicle launches is the absence of each launcher's 380-foot (116-meter) tall umbilical tower with its nine swing arms and large crane. The upper portions of the umbilical towers were dismantled and installed at the pads to serve as fixed service structures. A rotating service structure was built at each pad to provide an environmentally controlled Payload Changeout Room for inserting vertically handled payloads in the orbiter payload bay. Mounted on semicircular rails, this structure swings away from the vehicle prior to launch to prevent heat and blast damage.

Another change in the mobile launcher platforms is the replacement of the one large hole formerly at the center of each platform with three smaller openings, which separately accommodate the liftoff flames and hot exhaust gases emitted from the orbiter's three-engine cluster and the two boosters.

In 1986, another new facility was activated in the Complex 39 Area. The Solid Rocket Booster Assembly and Refurbishment Facility became the refurbishment and subassembly site for non-propellant booster hardware—primarily the forward and aft assemblies—initially performed in the Vehicle Assembly Building. NASA's Marshall Space Flight Center in Huntsville manages the Assembly and Refurbishment Facility.

Completed aft skirt assemblies of solid rocket boosters are transported to the Rotation, Processing and Surge Facility, another Shuttle era facility. This set of Complex 39 buildings is also the receiving point for new or reloaded booster segments from Utah. The aft skirt assemblies are integrated with the booster aft segments, and along with other booster segments, are then transported to the Vehicle Assembly Building for stacking and integration with other flight-ready booster components.

A number of structures and buildings in the Industrial Area to the south and nearby Cape Canaveral Air Force Station also were modified for Space Shuttle operations. For example, the Cape-side Hangar AF first used during the Gemini program now serves as the Solid Rocket Booster Disassembly Facility, the receiving point for the two spent

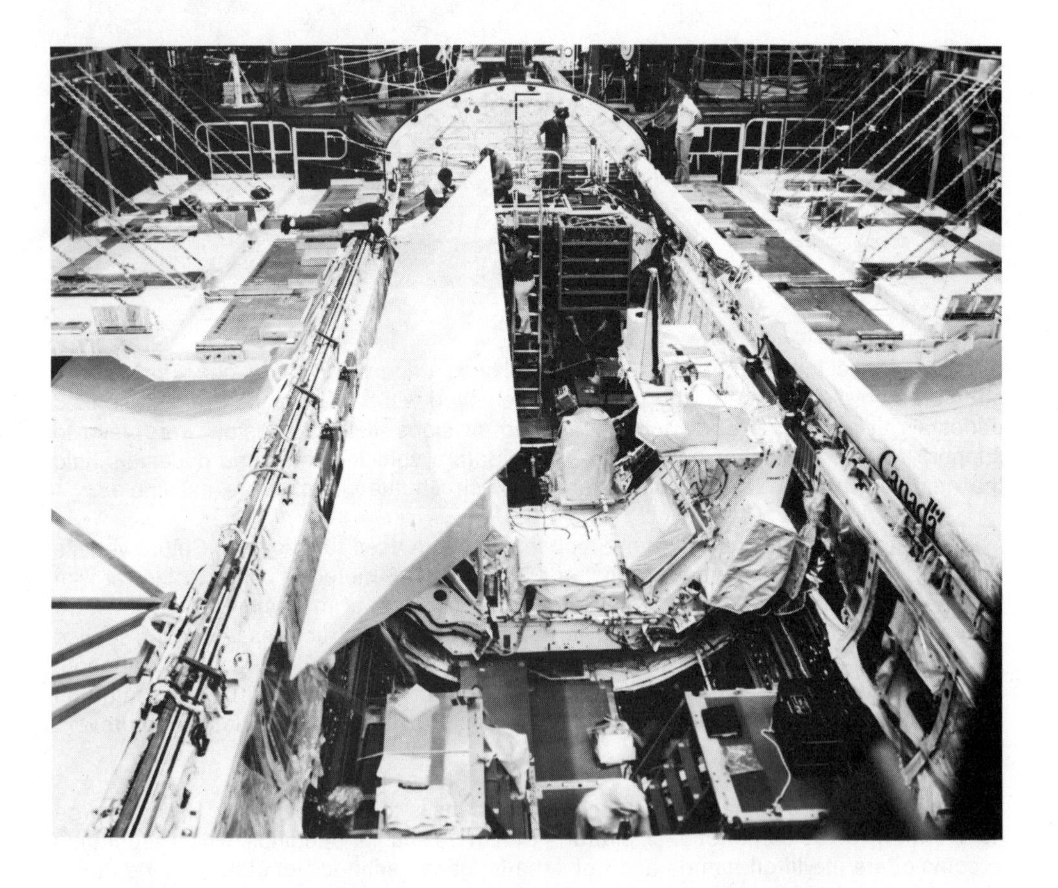

The Shuttle's first payload, processed in the Operations and Checkout Building and later installed in the vehicle in the Orbiter Processing Facility, was the OSTA-1 remote Earth sensing experiment, pictured here with the Canadian-built Remote Manipulator System shown along the payload bay edge.

A major repeat payload for Space Shuttle operational missions is Spacelab, a European-built scientific laboratory providing a shirt-sleeve environment in Earth orbit.

solid rocket boosters. After preliminary safing, the nose cone, frustum and aft skirt of each booster are removed and taken to Launch Complex 39 facilities for further processing.

The booster casings are disassembled into their major pieces and cleaned. The casings are then carried by truck to railroad cars and shipped to the manufacturer's plant in Utah, to be reloaded with propellant.

Some buildings and hangars at Cape Canaveral receive, assemble and check out payloads that are to be vertically integrated into the orbiter. These payloads, primarily automated satellites or spacecraft with attached stages, are then processed through one of two explosive-safe areas, also on the Cape.

These payloads are next brought over to the Vertical Processing Facility in the Industrial Area. This building, formerly used to encapsulate payloads in the nose cones of their expendable space vehicles, is now the site where two or more spacecraft and attached solid motors are integrated into a single Shuttle cargo package. Assembled payloads are inserted into a canister, which has interior dimensions that duplicate those of the orbiter's cargo bay. The canister is then sealed and transported to the pad. At the rotating service structure, the canister is hoisted and locked into position at the payload changeout room. The payload is then moved into the room, and the canister disconnected and lowered. The rotating service structure then swings around until it fits flush with the cargo bay of the orbiter. The payloads are then moved into the orbiter. Throughout this series of transfers, the payloads remain under "clean room" conditions.

Another Industrial Area facility converted for Shuttle operations is the Parachute Refurbishment Facility. This building was originally used to process parachutes for the Gemini manned space program, and was also used for a time as the KSC News Center during the Apollo program. Now it is the processing facility for the parachutes which slow the solid rocket boosters' descent into the Atlantic Ocean. The parachutes are washed, dried and prepared for reuse, and, along with new ones, stored until needed.

The nearby Hypergolic Maintenance and Checkout Facility was modified to process and store components of the Shuttle orbital maneuvering system and reaction control system. A Launch Equipment Test Facility, located at the Marshall Center during the Apollo program, was moved to KSC and then modified for Shuttle testing.

The Operations and Checkout Building, also in the Industrial Area, was originally designed for the assembly and checkout of Apollo spacecraft modules. It was converted to process Shuttle payloads which are integrated in a horizontal mode. It is used primarily for processing Spacelab, a scientific laboratory built by the European Space Agency (ESA). Spacelab and other payloads of this type are installed in the orbiter's cargo bay in the Orbiter Processing Facility.

The reusable Spacelab was developed as a modular concept so that its configuration can be varied according to specific mission requirements. Enough modules exist to keep two complete Spacelabs in continuous flow. One of a Spacelab's two principal components is a pressurized module which provides a laboratory where experimenters can work in ordinary clothing. The module is segmented to permit additional flexibility in size. It is connected to the orbiter's pressurized cabin by a tunnel. Payload specialists for Spacelab operations eat and sleep in the orbiter's cabin throughout the mission. Each Spacelab module is designed for at least 50 trips into space.

The second major element of a Spacelab is an open pallet that exposes materials and equipment directly to space. Five pallet segments are available, each 10 feet (three meters) long. The pallets are designed for large man-directed instruments that require direct exposure to space or broad fields of view. Such equipment includes telescopes, antennas and various types of sensors.

A Spacelab is transported into orbit by the Space Shuttle for missions lasting seven to 10 days. It remains inside the cargo bay of the orbiter throughout that time.

After a mission, the Spacelab is removed from the orbiter's cargo bay and returned to the Operations and Checkout Building. The experiments are removed for detailed analysis, a preliminary analysis having been done in flight. The Spacelab is then refurbished and readied for its next mission. In the interval, a second Spacelab may be launched if required.

ESA, under terms of an agreement with NASA, assumed responsibility for designing, developing, manufacturing, testing and delivering to NASA a Spacelab engineering model and a flight unit, plus ground support equipment and spare parts. NASA agreed to purchase one additional Spacelab and more if needed. Nations cooperating in the Spacelab program are: West Germany, Italy, France, the United Kingdom, Belgium, Spain, The Netherlands, Denmark, Switzerland, and Austria. NASA, which worked closely with ESA during all phases of Spacelab's development and production, has the responsibility for Spacelab operations.

In addition to the 10 Western European nations participating in the Spacelab project, the international flavor of the Shuttle program is further demonstrated by the orbiter's payload handling system. This system was designed and produced by a Canadian industrial team. The Remote Manipulator System deploys and retrieves payloads in orbit, using a 50-foot (15-meter) hinged arm—sometimes called the "Canadarm"—attached to the front of the payload bay. A second arm can be installed if needed. Each arm has remotely controlled television cameras, and lights that provide side viewing and depth perception.

The selection of potential new astronauts, the completion of initial flight tests of the orbiter, the production of other Shuttle elements and support equipment, the modification and construction of facilities and the development of a management team—all heralded the start of a new era in space operations.

This era began some two decades after the first U.S. venture into space—the launch of the 30.8-pound (14-kilogram) satellite, Explorer 1, on Jan. 31, 1958. Since then, unmanned spacecraft have probed the near and far reaches of space. Manned spacecraft have explored the lunar surface as well as expanded the existing knowledge of the Earth, the Sun, and the adaptability of people to extended flight in Earth orbit. The wealth of experience and knowledge gained from the accomplishments led to the development of the Shuttle.

With its versatility and reusability, the Space Shuttle is expected to open wide the doors to long duration exploration of space. A major step along this new path was taken April 12, 1981, with the first launch of the Space Shuttle.

Primary and backup flight crews for the first Space Shuttle flight visited KSC in January 1981 to train in emergency escape procedures from the launch pad. Wearing flight suits and carrying equipment that would be used during a mission made the training much more realistic. STS-1 Pilot Robert L. Crippen (in sunglasses) and Commander John W. Young (front, left) review procedures with their backup crew, Richard H. Truly (in helmet) and Joseph H. Engle.

chapter 13

FLIGHT TESTING the SPACE SHUTTLE

Perhaps no other mission during two decades of manned space flight was as critical to the future of the space program as the first flight of the Space Shuttle.

Columbia—first orbiter off the production line configured for flight in space—represented the first time in America's manned space program that NASA had undertaken an open-ended project. There was no longer just a single goal to be achieved.

The Space Shuttle was designed from the beginning as a key link in the exploration and development of the space frontier. The reusable capability it promised would allow the United States to deliver large payloads into Earth orbit, to repair spacecraft already in orbit, and, when required, bring payloads back to Earth. The Space Transportation System (STS), of which the Space Shuttle was the first major part, was expected to mature and evolve in ways hardly imagined at its beginnings.

Orbiter Columbia arrives at KSC, riding piggyback on NASA's specially modified 747. The spacecraft's next flight would be atop a column of flames several times its length.

But first the Space Shuttle had to prove itself. Columbia, the only orbiter that would be available for many months, was chosen for the complete flight test program of four missions. The three newer orbiters would fly as proven vehicles on their first launches.

Orbiter Columbia arrived at KSC from the manufacturing plant in California on March 24, 1979. It rode piggyback on a specially modified 747 jet. The 15,000-foot (4,572-meter) long Shuttle Landing Facility runway had been built to welcome an orbiter when it returned from space. But for the first two years these orbiters would descend from an altitude no higher than the back of a 747.

The next day, Columbia was towed to the new hangarlike building especially designed to handle this new breed of spaceship, the Orbiter Processing Facility.

The ensuing effort to ready Columbia for its maiden launch posed an unparalleled challenge to the KSC team. Technical difficulties had caused production delays, and there was still much work to be done on this complex new vehicle.

Columbia is lifted clear of its carrier aircraft by the Mate/Demate Device at the Shuttle Landing Facility. On its own wheels, the orbiter will be towed to its processing facility.

For protection from the burning heat of re-entry, the orbiter used a new system of reusable lightweight thermal tiles, made of ceramic-coated silica. They were superb insulators, but they were fragile and difficult to attach to the orbiter. Columbia arrived at KSC short some 8,000 of the needed 31,000 tiles. To make matters worse, new tests revealed that the bonding to the orbiter of many of the existing tiles might not withstand the stresses of flight. A decision was made to pull-test virtually all remaining tiles, with the result that most were removed, often strengthened, and reattached using improved methods.

Tile work became a major manufacturing task which had not been anticipated. The second high bay of the Orbiter Processing Facility was converted into a makeshift production shop to support the extensive tile work required on Columbia.

Meanwhile, other components for the first Space Shuttle were arriving at KSC during the summer of 1979, the beginning of a production flow that would grow into a steady stream of flight hardware. The external tank for the first mission (STS-1) arrived at the turning basin by barge on July 6. Four days later, the first of Columbia's three main engines arrived from the Stennis Space Center in Mississippi, where they had received their final flight acceptance tests. The remaining two were delivered in the weeks following.

Solid rocket motor segments produced by Thiokol Corp. began arriving by rail in mid-September, and by early December 1979, the segments were being "stacked" on a mobile launcher platform in the Vehicle Assembly Building. The assembly of the first Space Shuttle that would fly in space had begun.

By the time Columbia arrived at KSC in 1979, the nation's spaceport was a different place from what it had been in the days of Apollo. The NASA/ contractor workforce was more streamlined, trimmed to about half the size it reached during the peak of the Apollo Program. Richard G. Smith, a 19-year space program veteran who had served as deputy director of the Marshall Space Flight Center and NASA deputy associate administrator for space transportation systems at NASA Headquarters, became KSC's third Center director. He succeeded Lee Scherer, who retired from NASA after assuming the job four years earlier on the retirement of Kurt Debus.

Workers apply Thermal Protection System tiles to the sides of Orbiter Columbia. Despite apparent uniformity of many of the tiles, each is individually made for one location.

The dawning of the Shuttle era also saw a new industry team working at KSC. Rockwell International, manufacturer of the orbiters, was responsible for completing the work on Columbia and performing systems checks. Martin Marietta built the external tanks at NASA facilities in Michoud, La., then did finishing work and helped launch them at KSC. United Space Boosters assembled the two solid rocket boosters. They also were responsible for recovering the spent casings after they parachuted into the ocean off Cape Canaveral, and refurbishing all but the propellant segments for another launch.

On Nov. 24, 1980, Columbia emerged from 20 months in the Orbiter Processing Facility. The orbiter was towed on its own wheels over to the VAB, initiating a round-the-clock launch processing schedule aimed at achieving the first flight by the end of March 1981.

During the next few days technicians mated the orbiter with its external tank and solid rocket boosters, already assembled on a mobile launcher platform in High Bay 3. And at 8 a.m. on Dec. 29, about 16 million pounds (7.3 million kilograms) of Space Shuttle, platform and crawler-transporter inched out the giant doors and rolled slowly toward Pad A. By that evening Columbia rested on its launch pad pedestals, bathed in floodlights and poised for its first journey into space.

Columbia still had a few hurdles to overcome before that journey could begin. NASA had never before launched a new manned space vehicle with a crew aboard on its first flight, but that was to be Columbia's mission. Program managers decided to add a new test to the checkout and launch procedures, one designed to raise confidence in the complex, state of the art orbiter main engines. They were to be fired on the pad for 20 seconds, another first in any NASA manned program.

The true size of the Space Shuttle external tank can be seen when human beings stand next to it. This early tank is in the transfer aisle of the Vehicle Assembly Building.

On Feb. 20, 1981, the engines roared to life at the end of a 72-hour countdown demonstration test that served as a launch rehearsal. They shut down as planned. After reviewing the data, program officials declared the firing a success.

The final major test still between Columbia and its first voyage into orbit was a "dry" countdown demonstration test—a full-dress rehearsal for the astronauts and the launch team. The test went as planned, but a tragic accident claimed the lives of two Rockwell technicians when they entered Columbia's aft engine compartment while it was still being purged with gaseous nitrogen.

Two ships--"NASA's Navy"--were specially designed and constructed to retrieve the spent solid rocket booster casings. Preparation for the first Shuttle launch included intensive practice using dummy boosters. The ships were built by Atlantic Marine Shipyard at Fort George Island near Jacksonville, Fla., and leased by NASA.

KSC was stunned by the tragedy. But preparations for the launch continued, and April 10 was set as a firm launch date.

The moment that KSC had been anticipating for a decade was rapidly approaching.

The rollout milestone is achieved. Columbia, mated with its external tank and solid rocket boosters, inches its way from the Vehicle Assembly Building to Launch Pad 39A.

NASA had selected veteran astronaut John Young—who had flown aboard two Gemini and two Apollo missions—as commander for the first Space Shuttle flight. Astronaut Robert Crippen, who had not yet flown in space, was named pilot. There would be no real payload. Instead, Columbia was fitted out with an extensive development instrumentation package, designed to measure its performance and the stresses and strains encountered.

The STS-1 crew had practiced extensively on flight simulators at Johnson Space Center, and had made many landings with a Gulfstream jet modified to have the handling and landing characteristics of an orbiter. They landed this plane many times at both KSC and the extra-long runways at Edwards AFB in California, the primary landing site. They were as ready and experienced as any activity short of flying a Space Shuttle could make them.

The weather was perfect on Friday morning, April 10. The crew entered Columbia at a little after 4 a.m., ready for a planned 6:50 liftoff. An estimated 80,000 guests crowded the main thoroughfares on the Center, the NASA Causeway and Kennedy Parkway. More than a hundred busloads of VIPs had gathered at the NASA guest site. More than 2,700 members of the news media were working at the KSC News Center; many had been there for days. And off the Center, hundreds of thousands of people lined the roads and beaches surrounding KSC, anxiously waiting to witness the historic event.

Excitement was reaching a fever pitch at T minus 9 minutes, when the countdown came out of its last planned hold. And then a major problem emerged. The backup computer on Columbia was not properly synchronizing with the four operating flight computers. They could not exchange data.

Liftoff was rescheduled for 10:20 a.m., while the problem was investigated. But then bad news came over the loudspeakers; the launch would have to be postponed.

Computer experts isolated the problem Friday night, and on Saturday the countdown resumed. Liftoff was rescheduled for 7 a.m. Sunday, April 12. But the weather was now doubtful. The launch team decided to proceed, and to make a decision on weather when that became necessary.

By Saturday night the weather looked good. The countdown proceeded toward liftoff without a hitch. Once again the guests and spectators crowded into KSC, and gathered at good viewing spots around the Center. And this time they were not disappointed. At a fraction after 7 a.m., the orbiter main engines roared to life, followed seconds later by a giant cloud of flame from the two solid rocket boosters. Columbia lifted majestically off the pad and arched out over the Atlantic, heading into the golden dawn and a new age in space for the United States and the world.

* * * * * *

At 10 a.m. EST on Nov. 12, 1981, Columbia lifted off Pad A a second time. It had been a long seven months between flights, far longer than NASA liked. Still, the time spent in the Orbiter Processing Facility had been shortened from 610 days for STS-1 to 103 days. Columbia had required still more tile replacement and repair, and some launch pad modifications were needed to cushion the overpressure wave that followed the ignition of the solid rocket boosters.

STS-2 also had a payload, in addition to the development flight instrumentation which had been carried on STS-1. The package of scientific instruments was called OSTA-1, for the former NASA Office of Space and Terrestrial Applications. The space application areas under investigation were remote sensing of land resources, environmental quality, ocean conditions and meteorological phenomena.

The largest instrument in the package, a side-looking radar called the Shuttle Imaging Radar-A (SIR-A), provided one of the serendipitous dividends for which the space program is famous. While operating over a remote area of Egypt, one of the driest deserts in the world, this radar penetrated the loose sand to a depth of up to 6.6 feet (two meters). Hidden beneath the loose, flowing sands were ancient watercourses, riverbanks, dried streambeds, etc., as they must have looked hundreds of thousands of years in the past.

Although it worked only where the sand covering was very dry, this surprising ability of the SIR-A radar was noted and targeted for future missions. The other instruments also performed well.

STS-2 also featured the first flight of a device certain to play a very important part in the future operation of the Space Shuttle, the Canadian-built Remote Manipulator System. This 50-foot (15.2-meter) long Canadarm, with television cameras mounted at wrist and elbow, would have an important role to play on many future missions. On this mission the Canadarm was tested in all its modes, but not under heavy loads.

The second STS mission had to be curtailed due to a problem with one of the

The Canadian-built Remote Manipulator System, sometimes called the Canadarm, moved a magnetometer device around the Columbia in orbit, checking for the presence and strength of magnetic fields on the third Space Shuttle flight.

fuel cells. Although this situation was not dangerous, mission controllers conservatively decided to shorten the flight. It landed after two days and 36 orbits, touching down on the long runway at Edwards AFB. The crew of Commander Joe Engle and Pilot Richard Truly reported that overall it had been a good mission, and they had accomplished 90 percent of their objectives despite having the planned flight time cut from five days to two.

* * * * * *

KSC was working toward a projected goal of up to 24 launches a year, with an interim goal of half that figure, or a launch a month. This could only be achieved when three or more orbiters were available. In the meantime, the ability to produce external tanks and solid rocket booster components was slowly but steadily increasing, with the eventual aim of being able to support the maximum anticipated flight demand. But completing the flight test program and having an operational Space Shuttle available was the first priority.

STS-3 lifted off at 11 a.m. EST on March 22, 1982, with a third two-man crew, Commander Jack Lousma and Pilot Gordon Fullerton. The payload was another scientific package, the OSS-1, named after the NASA Office of Space Science. These instruments were mounted on a pallet designed for use with the Spacelab system. The mission was intended to last seven days, and not only achieved that goal but was extended by one day. The planned landing site at Edwards AFB was too wet to be used, and the backup site at Northrup Strip in New Mexico had high winds on the seventh day. The mission was stretched by 24 hours, to eight days and 129 orbits, by which time the winds had died away and a safe landing was achieved at Northrup.

In orbit Columbia was rotated and held in several different attitudes in relation to the Sun, to test heat dispersion throughout the vehicle. Between tests the vehicle was rolled to prevent excessive heat buildup in any one area. The Canadarm was also exercised again, this time with a load.

This mission also featured the first Shuttle Student Involvement Project (SSIP) to fly, in the middeck area. It was originated and developed by a high school student, with commercial sponsorship. The middeck area also had the first small experimental Continuous Flow Electrophoresis System (CFES), to produce high-value drugs at low cost, and a Monodisperse Latex Reactor. The latter could produce micron-sized particles of uniform diameter that have a wide variety of scientific and industrial uses. Both systems had the potential to provide substances that, pound for pound, would be far more valuable than gold. And the CFES had the distinction of being the first Space Shuttle payload developed by and for private business, for future commercialization and profit. It was a joint venture between McDonnell Douglas and Johnson & Johnson.

Columbia also flew more development flight instrumentation, as it would for several future missions. But by the end of this eight-day flight, the ability of the orbiter to sustain long missions had been clearly established. Its ability to land outside its normal home ground of California and Florida had been proven. One more development flight was scheduled, and if all went well, the Space Shuttle would be declared operational.

In the Launch Control Center firing room, a vigilant launch team counts down to another successful liftoff.

The last development flight for the Space Shuttle was also the first for the customer—the Department of Defense. Commander Thomas Mattingly and Pilot Henry Hartsfield, both military officers on detached duty with NASA, would oversee the first "mixed" payload of military and civilian cargo.

STS-4 lifted off at 11 a.m. EDT on June 27, 1982. This was the first STS mission to be launched on time and with no delays in schedule. Columbia had remained in the Orbiter Processing Facility only 42 days. The processing routine was now becoming well-established, and experienced crews were able to do the work more quickly.

In addition to the classified Air Force payload in the cargo bay, STS-4 also carried the second CFES and Latex Reactor experiments in the middeck area. Nine small experiments were contained in the first "Getaway Special," a low-cost cargo container designed especially for economical projects to be carried in the cargo bay. These had been prepared by students at Utah State University. The crew also operated the Canadarm again, took medical measurements on themselves, and used hand-held cameras to take photographs of the cloud cover below.

The only major problem encountered was not with the orbiter. The two spent solid rocket boosters, which had been successfully recovered on all three earlier flights, were lost when their parachutes failed to deploy properly. They hit the water with such force that both sank. They were later examined by an underwater remote camera, and officials decided that recovering them was not worth the cost. The problem that had caused the parachute failure was located and corrected without a close examination.

The landing at Edwards AFB, after a mission duration of seven days, was on a concrete runway the same length as the one at KSC. President and Mrs. Ronald Reagan attended the landing and welcoming ceremonies.

The third Space Shuttle flight lifts off with Columbia from Pad 39A, at 11 a.m. EST on March 22, 1982. This was the first use of the external tank without paint over its coating of insulation.

Clouds of steam and smoke billowed as the first Space Shuttle plunged away from the launch pad, atop a huge lance of flame from the solid rocket boosters' nozzles.

chapter 14

The WORKING SPACE SHUTTLE

Getting those first four development flights "under the NASA belt" had been a long, tough job. The extensive delays had caused a backlog of payloads, more than the Space Shuttle could reasonably expect to handle in the near future. Some commercial communications satellites originally on the Shuttle manifest had already shifted to Ariane, a lively new unmanned competitor built by the European Space Agency. Deltas and Atlas-Centaurs, scheduled to be deactivated by this time, were still being launched, although on reduced schedules. The lower cost of a Space Shuttle flight made this vehicle very attractive to potential customers, but there was a long waiting period for cargo space. For the moment, there were payloads enough for all four vehicles.

NASA still faced the formidable task of working up to a launch schedule of two Shuttles a month. That could only be done when the full fleet of four authorized orbiters were all flying. Challenger was to be the second one off the production line, followed by Discovery and then Atlantis. But Columbia had to fly the next mission because it was still the only orbiter available.

STS-5 lifted off at 7:19 EST on Nov. 11, 1982. This first paying mission carried two commercial communications satellites, a Satellite Business Systems-3 and Anik-3, both Hughes HS-376 spacecraft. Each was also the next satellite in line in a series, their predecessors having been launched on unmanned vehicles. And there would be more in both lines to follow.

This first Space Shuttle launch of two communications satellites was also symbolic of the future. The geosynchronous communications satellite had spawned a new industry—telecommunications from space—and it was generating two billion dollars a year in revenue. No other civilian application of space technology had caught on so quickly, or been so rapidly converted into moneymaking commercial ventures. And with an already thriving business expected to benefit from the less expensive launch fees promised by the Space Shuttle, future growth seemed unlimited.

Satellite Business Systems soon paid to have its fourth satellite launched by the Space Shuttle, and Telesat of Canada sent up both an Anik C-2 and Anik D-2. Indonesia decided to augment or replace aging satellites in its Palapa domestic communications system. A Shuttle took up Palapa B and Palapa B2. Western Union, the first American company to have its own domestic communications satellites, sent up Westar VI. American Telephone & Telegraph sent up Telstar 3-C. India sent up Insat 1B, which was an unusual combination spacecraft that could photograph the Earth every

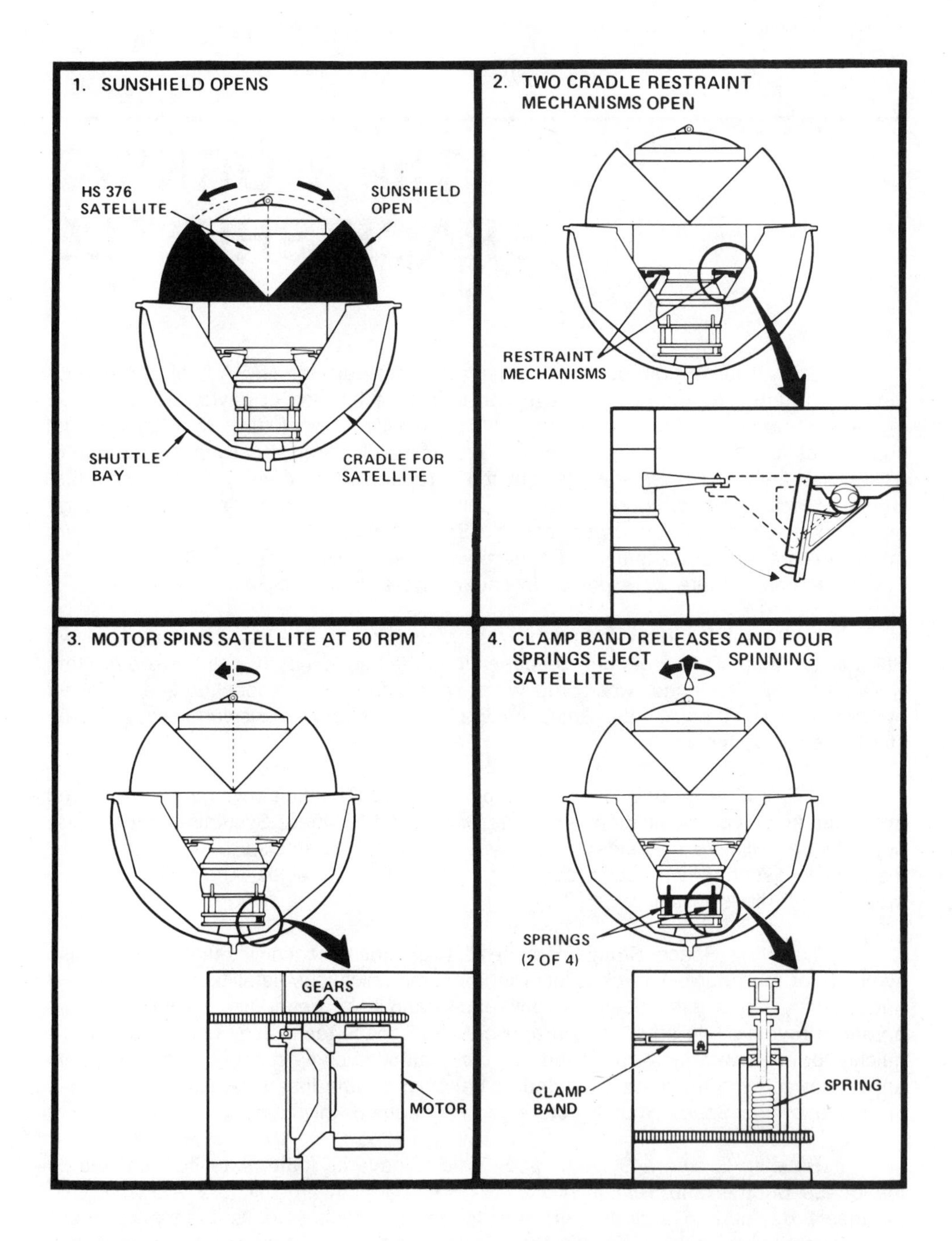

The most frequently launched Space Shuttle payloads have been communications satellites, usually carried into orbit in special cradles specifically designed to protect them until they are released.

30 minutes as a weather satellite, relay television signals direct to small antennas like a direct broadcast satellite, and also serve as a general telecommunications spacecraft.

Hughes Aircraft, which had built more satellites for other firms than any American company, decided to enter the burgeoning field of commercial satellite communications. Hughes established a domestic system of three spacecraft called Galaxy, where the design was based on its proven HS-376 model. Their business plan required transponders to be sold in orbit to individual buyers, who might or might not be regulated common carriers. This allowed Hughes itself to avoid becoming a common carrier. All three Galaxy spacecraft were launched on Delta vehicles. Hughes also built a new spacecraft too large for the Delta called Syncom IV, for which it kept ownership but leased the entire capacity to the U.S. Navy. Two Syncom IVs were launched in 1984 and two more in 1985.

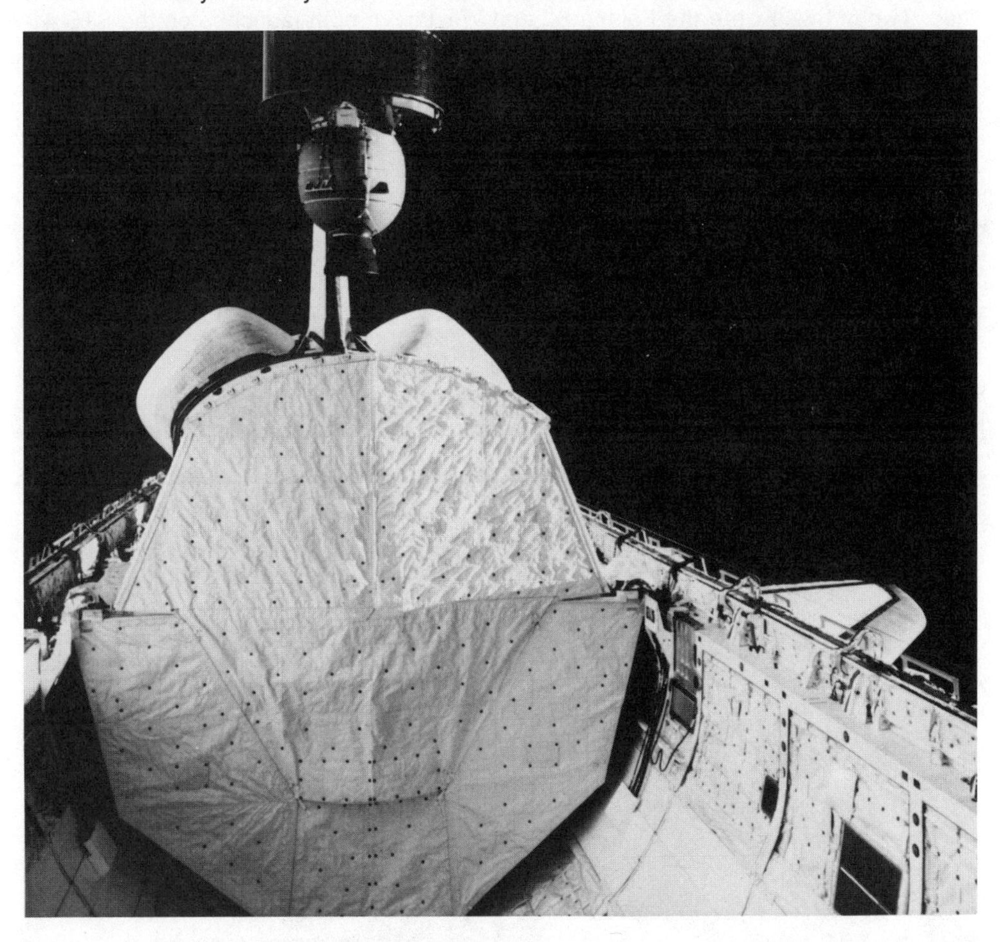

Turning at 50 revolutions per minute, the Anik C-3 spacecraft rises out of the cargo bay on the STS-5 mission. It was housed in a sun shield identical to the one in the foreground (now closed) that housed the SBS-3 spacecraft.

NASA launched the largest communications satellite of them all for itself. This was the first Tracking and Data Relay Satellite, or TDRS. This massive spacecraft was designed to become a tracking station in the sky, capable of replacing most of the network of ground stations that NASA had for many years maintained around the world. With two TDRS satellites in geosynchronous orbit and on opposite sides of the world, a Shuttle orbiter could maintain constant communication with the ground for 85 percent of its flight time. With the ground stations, the average was 15 percent. And the satellites, once built and launched, were far less costly to maintain than the ground stations they replaced.

Many communications spacecraft were successfully deployed from the orbiter, but three failed to reach geosynchronous altitude when their attached rocket motors were fired. Two of these were the Palapa B2 and Westar VI, both launched on the STS 41-B mission. Their Propulsion Assist Module rocket motors failed shortly after ignition. Left in stable but low orbits, both spacecraft were retrieved during mission STS 51-A in November 1984 and brought back to Earth for refurbishment. The third was TDRS, which was being boosted by the new Inertial Upper Stage. Fortunately, the rocket engine burned long enough to carry TDRS halfway to its planned 22,300-mile (35,889-kilometer) geosynchronous altitude. And there was enough extra onboard attitude control propellant to allow its own small thrusters to gradually work the huge satellite up another 11,000 miles (17,703 kilometers). Over a period of several months, NASA controllers at Goddard slowly and carefully raised TDRS, using many short burns to prevent overheating the small rockets. It eventually reached its planned station, and went to work.

TDRS had cost about 100 million dollars. It also had the potential to save NASA much more than that over the next 10 years. Saving TDRS was thus a double financial benefit to NASA. TDRS also was badly needed to support the upcoming Spacelab flight.

Payload Specialists Ulf Merbold, left, and Byron Lichtenberg, back to camera, along with Mission Specialist Robert Parker, operate the complex equipment inside the Spacelab 1 module on the STS-9 mission.

Only this large and complex satellite had the capacity to properly relay to ground stations the huge amounts of data that Spacelab would generate.

* * * * * *

The second major area of Space Shuttle utilization was scientific experimentation, using human beings to operate the instruments in orbit instead of automated spacecraft or ground control. By far the largest and most complex science program was that of Spacelab, which was scheduled to eventually fly two or three times a year. But smaller and less complicated instruments would fly on virtually every Shuttle flight, performing a wide variety of experiments and observations.

It was sometimes difficult to distinguish between true scientific exploration of the unknown, and the development of new or improved products through in-orbit work in materials technology. There were also many experiments in the latter category, often flying side by side with science experiments in the Spacelab or on a pallet. Their primary purpose was to take advantage of the high vacuum and microgravity of space to produce unique or highly valuable products for eventual sale in commercial markets.

Spacelab 1 flew on the ninth Shuttle mission, and was an immense success. The first payload was deliberately designed to cover the five major areas of scientific inquiry and technical development, to each of which an entire Spacelab might be devoted in the future. The primary purpose was to demonstrate what could be done in each area, rather than attempt to do extensive research or experimentation. The science categories were atmospheric physics and Earth observation, space plasma physics, astronomy and solar physics, and life sciences. Materials science and technology was the fifth area, where development work could be done to provide useful commercial products, based on what had already been learned about the unique manufacturing conditions available in space.

Spacelab also featured the flight of two non-NASA astronauts, payload specialists who had special training in operating the Spacelab instruments but only a minimum of the rigorous training required of NASA astronauts. One of these was from the European Space Agency, builder of the Spacelab and operator of half of the experiments aboard. The second was a researcher sent by the Massachusetts Institute of Technology. An unusually capable ground communications system enabled the operating crews in orbit to perform some work under close direction of the primary scientific investigators on Earth. To completely analyze the immense amounts of data Spacelab 1 obtained was a task of several years duration.

The third Shuttle development flight had introduced a new type of experiment into orbit. The Shuttle Student Involvement Program experiments became a continuing payload on many succeeding launches. Another category, the Getaway Specials, were carried in sturdy containers in the cargo bay, and for these there was a reduced charge. They were intended for students at the college level who could obtain sponsors willing to foot the bill, or for individuals or companies who had small experiments they wanted to see performed in orbit.

Another large though less active science project was the reusable Long Duration Exposure Facility (LDEF), sponsored by NASA. This huge cylinder, some 30 feet (9.1

meters) long and 14 feet (4.3 meters) wide, was carried into space to be left for a year or more. The body of LDEF consisted primarily of some 86 trays, which accommodated more than 50 experiments (some being too large to fit into a single tray). These were largely passive in nature, with few or no requirements for power and movement. One of these was a huge number of tomato seeds, which were to be distributed to school children and planted after having been in space. They were to be compared to plants grown at the same time from seeds not exposed to space, to see if any differences appeared in the mature plants.

NASA also deployed a scientific spacecraft of its own, the Earth Radiation Budget Satellite, or ERBS. Its purpose was to measure the incoming heat from the Sun and its re-radiation into space, to learn more about how the Earth managed to avoid heating up from the constant intake of sunlight.

Astronaut Bruce McCandless takes the first ride at the end of the Canadarm on the STS 41-B mission, after the launch of the Westar VI and Palapa 2-B satellites. He is above the empty Westar VI sun shield.

NASA continued the work started during previous manned space flight programs on how humans work and function in space, including attempting to discover the cause and cure of "space sickness." This temporary and sometimes debilitating illness strikes about half the members of astronaut crews during the first days in weightlessness. A continuing series of experiments were performed over many flights. The life sciences experiments on the Spacelab 1 mission provided extensive data on human reactions in space, forming a data base against which other reactions could be measured. Although not a major problem, space sickness is an annoyance which astronauts would rather not have to endure.

* * * * * *

From the beginning of space flight, the promise of unique materials and products that could be produced in microgravity had intrigued chemists and engineers. With the Space Shuttle the ability to return to space again and again became a reality, making it practical to plan a continuing series of developmental steps leading to an operational capability. Several forward-looking companies already had agreements with NASA, and more were pending. One of the earliest was McDonnell Douglas, working in association with the Ortho Pharmaceutical Corp.

The first experimental Continuous Flow Electrophoresis System had flown on STS-3. This was a biological separation device that, in microgravity, could produce drugs of such purity and strength that their value far exceeded the expense of producing them in space. Manufacturing these drugs on Earth, in the grip of gravity, was possible, but the quantities produced were small and prohibitively expensive. High-volume production in microgravity could lower the price and make the drug more widely available, thus benefiting untold numbers of people, while still making a good profit for McDonnell Douglas

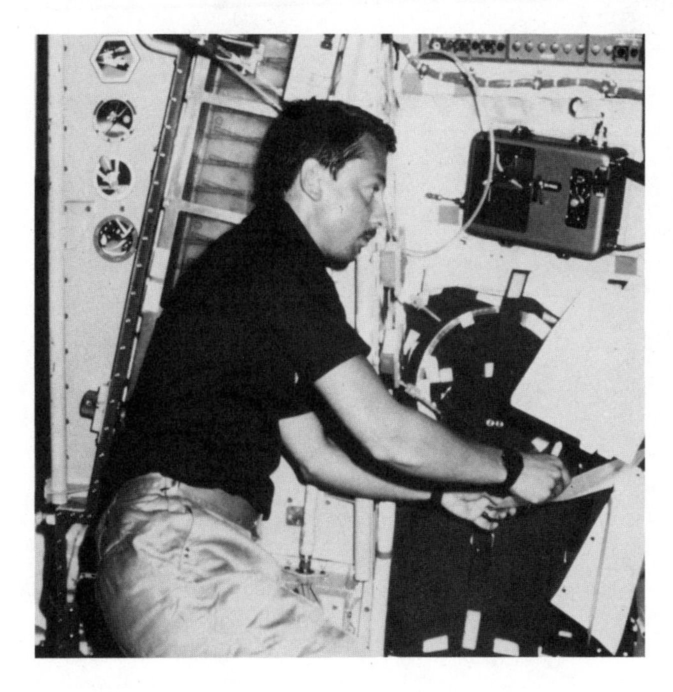

Charles Walker, the McDonnell Douglas engineer who became the first payload specialist to operate his firm's equipment in orbit, logs some flow rate calibrations on the Continuous Flow Electrophoresis System on the STS 41-D mission.

and Ortho. This was the type of manufacturing in space that NASA wanted to encourage.

After several flight tests, McDonnell Douglas sent up a larger version of its machine on the 12th Shuttle flight, along with its own operator, engineer Charles Walker. Walker thus become the first commercial payload specialist. Although some mechanical problems were encountered, Walker resolved them in time to produce a large quantity of drugs and prove that the concept was practical and workable. More flights were set for the future.

Strangely enough, the first product actually made in space and sold for money was produced by NASA. The Monodisperse Latex Reactor, which had first flown on STS-3 along with the CFES, on later flights produced a quantity of tiny, highly uniform latex beads that had a variety of scientific and industrial uses. Beads of this almost microscopic size that were also uniform could be used to calibrate scientific and measurement devices that were very difficult to certify any other way. Beads that size produced in gravity could not be made uniform enough to be useful. Sold in small packets, the beads brought in enough money to partially defray the costs of the several experiments.

Astronaut Dale Gardner approaches the slowly turning Westar VI satellite with grappling device extended. After the device was locked in place, the Canadarm, right, grasped it and placed the satellite in the cargo bay.

Astronaut Dale Gardner, left, holds a "For Sale" sign to indicate the successful recapture of one of two "lost" communications satellites from the STS 41-B mission. Riding the end of the Canadarm is teammate Joseph Allen.

The 3M company also made a long-term development agreement with NASA, and flew a set of experiments on the 14th mission. It dealt with the diffusive mixing of organic solutions, and was the first of more than 70 planned organic and polymer science experiments. 3M later reported the tests had been highly successful, although the equipment was produced in about one-third the time normally required to prepare a flight experiment.

West Germany, which had the bulk of the materials experiments on Spacelab 1, made reservations for 25 Getaway Specials for similar investigations. This member of the European Space Agency also independently decided to take an entire Spacelab flight for an extensive materials science investigation.

* * * * * *

A new area that NASA pioneered during the early years of the working Space Shuttle was the recovery and refurbishment of failed satellites. Although many spacecraft in orbit and on interplanetary trips had failed and been restored by ground control, no satellite had ever been physically repaired in space, or recovered and brought to the ground for repairs. The Shuttle did both within a single year, 1984.

Solar Maximum Mission, a NASA scientific satellite to study the Sun, had failed after a short operational life. Ground controllers felt certain it could be repaired by replacing one electronic box and a few components. The crew that placed the Long Duration Exposure Facility in orbit went on from there to rendezvous with Solar Max. It was captured and brought into the cargo bay, where two astronauts wearing space suits replaced the needed box and installed new components. The satellite was tested and released back into orbit, where it functioned for another five years before re-entering the Earth's atmosphere.

The WESTAR VI and Palapa B-2 communications spacecraft that had been left in low orbits by the failures of their attached booster stages were recovered and returned to Earth. This was possible because the orbits they had reached were low enough that ground controllers could bring them back down within range of the orbiter. The rendezvous proved much easier than wrestling the two satellites into the cargo bay and securing them there, but two astronauts in space suits eventually succeeded. Once back in the Hughes factory, the two spacecraft were examined, and proved little the worse for wear. Both were refurbished and prepared for a second try at useful life in orbit.

In between these two missions NASA performed a test which has a high potential for future applications—refueling satellites in orbit. Refueling gear and hydrazine tanks were carried up on STS 41-G in the cargo bay. Astronauts in space suits demonstrated that it was possible to attach hoses and valves to exhausted satellites and pump hydrazine into their dry tanks. Since using up their fuel was the most common reason satellites were "put out to pasture" (placed on the inactive list), the experiment proved it was possible to revive those the Shuttle could reach.

The ability to repair and refuel satellites in orbit, or recover and return some to the ground, opened up a whole new chapter in the utilization of space. And all three tasks were accomplished as secondary objectives, after the main business of each mission had been completed. The Space Shuttle and its crews were steadily expanding their horizons.

Most Space Shuttle missions lasted from six to 10 days, and the size and capacity of the vehicle made it possible to carry a large crew. Six-person crews became common, seven were flown when necessary, and the STS 61-A Spacelab D-1 mission had a crew of eight. Crews that large, with a week in orbit, could accomplish a great deal of work. There was time for many smaller projects along with the major ones.

The Long Duration Exposure Facility hangs in the microgravity of space, held by the Canadarm. It was later released into low Earth orbit. The STS 41-C crew then went on to capture and repair the Solar Max satellite.

Two of these smaller projects were photographic in nature. A group of observatories had banded together and arranged for a special camera, called Cinema-360, to be flown in the cargo bay and in the crew cabin on several missions. Its large lens, when operating in the totally air-free and extremely clear region of space, returned spectacular film suitable to be shown on planetarium domes. IMAX, another large-film camera system operated by the astronauts from inside the orbiter, did an equivalent job for showing on extra-large wall screens. The latter was available to more people, and shows appeared at several places. These included the KSC visitors center, Spaceport USA, and the Smithsonian's Air and Space Museum in Washington, D.C.

On one mission NASA deployed a solar cell wing out of the cargo bay, one equipped with only a few cells of different types. The experiment was designed to test the wing itself more than the cells. Folded, this 13-foot (4-meter) wide sheet stood only seven inches (18 centimeters) high in its container, but it erected to a height of 102 feet (31 meters). An operational version, fully covered with solar cells, could produce a great deal of power for future in-orbit applications.

The astronauts also proved the design of their extravehicular space suits in actual practice. After this was accomplished, they operated the Manned Maneuvering Unit, or MMU, a large device which the astronaut attaches to his body somewhat like an oversized backpack with arms. The MMU has built-in thrusters and a control system, among many other features, enabling an astronaut to maneuver freely away from the orbiter. The MMUs were used during all the retrieval and refurbishment operations, allowing astronauts to fly to and around the satellites, once the orbiter had brought them to the vicinity.

The astronaut crews performed many other experiments and miscellaneous tasks during the first operational years, doing varied and different kinds of jobs than had been attempted previously in the space environment. The Space Shuttle was proving that it possessed the versatility and multicapabilities that its designers had promised from the beginning.

A SYNCOM IV satellite spins slowly away into space from the Space Shuttle orbiter on the STS 41-D mission. Some 45 minutes later, at a safe distance, its onboard engine ignited to lift it to a higher orbit.

chapter 15

CHALLENGER: SETBACK and RECOVERY

On Sept. 29, 1988, at 11:37 a.m., the Space Shuttle Discovery proudly lifted off Pad 39B at the Kennedy Space Center. In outward appearance, the launch of mission STS-26 looked the same as many before it. The orbiter with its distinctive black and white banding, the cigar-shaped external tank and the twin solid rocket boosters—all seemed no different than 25 other Shuttles which had lifted off from Launch Complex 39.

Challenger crew members, left to right, Christa McAuliffe, Gregory Jarvis, Judy Resnik, Dick Scobee, Ronald McNair, Mike Smith and Ellison Onizuka stand in the White Room of the orbiter access arm at Pad 39B following a prelaunch test Jan. 8, 1986.

But if one listened closely to the cheering crowd on hand to witness the launch, it immediately became clear that this was no ordinary mission. Journalists watching the ascent maneuvers on television at the KSC News Center were especially vocal when the solid rocket boosters (SRBs) separated from the orbiter and external tank a little over two minutes into the flight. "What a beautiful sight!" one yelled. "Go, U.S.A., go!"

A close-up inspection would have revealed that the solids powering Discovery were a redesign of their predecessors, and had never flown before. The orbiter itself had undergone more than 200 modifications since the last Shuttle mission. Even the pad from which STS-26 was launched had new features and equipment.

The flight of Discovery symbolized the historic resumption of the American manned space flight program. Nearly three years before, on the bitter cold morning of Jan. 28, 1986, the Shuttle Challenger had blasted off on mission 51-L from the same pad as STS-26. But what seemingly began as another near-routine flight ended a brief 73 seconds later when the spaceship and her crew of seven were lost in a massive structural breakup. It was the most profound setback the U.S. manned space flight program had ever suffered, and a national tragedy. What happened between the Challenger accident, and the return to flight 32 months later with the launch of Discovery, is a story of painstaking re-examination and recovery.

Workers haul Challenger debris onto the deck of the Navy's USS Preserver (ARS-8), one of 16 surface ships involved in the extensive search and salvage operation.

Challenger was the third orbiter NASA ordered from Rockwell International and the second to fly into space. It was originally built as a structural test article and then modified for operational flight. OV-99, as Challenger was designated, became the workhorse of NASA's four-orbiter fleet, completing nine missions from April 1983 to January 1986.

CHALLENGER: SETBACK and RECOVERY

A number of "firsts" were achieved on Challenger flights: the first night launch and landing of the Shuttle program (STS-8), first landing of the Shuttle at Kennedy Space Center (STS-41B), first space walk of the Shuttle program (STS-6), the first American woman in space (STS-7), first flight of an African-American (STS-8) and first in-orbit satellite capture and repair operation (STS-41C).

STS 51-L was also going to be a historic first: the first teacher in space. Payload Specialist S. Christa McAuliffe was a New Hampshire schoolteacher chosen from thousands of applicants. She was scheduled to teach a class from space once the Shuttle achieved low Earth orbit.

Mission commander was Francis R. Scobee; Michael J. Smith was the pilot. Gregory B. Jarvis was the second payload specialist, and there were three mission specialists: Ronald E. McNair, Judith A. Resnik and Ellison S. Onizuka. Smith, McAuliffe and Jarvis were on their first Shuttle flight. Primary payload for the mission was a Tracking and Data Relay Satellite (TDRS). Challenger had carried aloft the first TDRS spacecraft on her maiden voyage three years earlier.

STS 51-L blasted off from the pad at 11:38 a.m., six days later than planned and two hours late on the rescheduled day. Later photographic reviews of the launch showed that at less than a second after ignition, a puff of smoke spurted from the area of the aft field joint on the right solid rocket booster. The presidentially appointed task force which investigated the loss concluded that it was the failure of this joint which culminated in the breakup of the vehicle some 73 seconds later.

Less than an hour after the accident, the most intense search and recovery operation ever undertaken at sea was already under way. The search for Challenger debris lasted seven months, and involved a combined operation of the United States Coast Guard, Air Force, Navy and NASA, as well as private vessels working for the government. Altogether 16 surface vessels, a submarine, four remotely operated vehicles and two manned submersibles, five aircraft and seven helicopters were used.

From depths exceeding 1,200 feet (366 meters) and covering an area greater than 486 square nautical miles (1,668 square kilometers), the salvage teams extracted 118 tons (107 metric tons) of Challenger debris. NASA ranked location and recovery of the suspect right solid rocket booster and the orbiter crew compartment as the top priorities.

After extensively searching for floating debris, the salvage teams used sonar to map and identify debris beneath the ocean's surface. Of the more than 870 sonar contacts made, about 700 were actually investigated. Through this meticulous process, searchers recovered the right SRB section showing where burn-through of the joint had occurred, as well as much of the orbiter compartment housing Challenger's crew. Almost half of the orbiter and 90 percent of the boosters were brought to the surface. The 215,000 pounds (97,524 kilograms) of debris were eventually placed in deactivated Minuteman missile facilities at Complexes 31 and 32 at Cape Canaveral Air Force Station.

Within hours of the accident, NASA began forming an interim task force to investigate its cause. This team, comprised of top management officials from Kennedy Space Center and other NASA locations, became the 51-L Interim Mishap Review Board.

A frustum from one of Challenger's two solid rocket boosters is lowered into its final storage site--two deactivated, underground Minuteman launch silos at Launch Complexes 31 and 32 on Cape Canaveral Air Force Station.

Jesse Moore, then NASA associate administrator for space flight, was named chairman. Three locations, one at KSC and two at the adjacent Cape Canaveral Air Force Station, were designated storage and reconstruction areas for salvaged Challenger debris. Experts from the National Transportation Safety Board, which investigates aircraft accidents, were called in to examine the physical wreckage as part of the effort to determine what had happened, and why.

The day after the accident, processing of the other three orbiters—Discovery, Atlantis and Columbia—was suspended, and mission planning placed on hold. Jesse Moore ordered that no new Shuttle-related contracts be signed. A photo team presented the investigation board with the first photographic indication of an anomaly, at or near the field joint between the aft and aft center segments of the right SRB.

On Feb. 3, President Ronald Reagan signed an executive order creating the Presidential Commission on the Space Shuttle Challenger Accident. Former Secretary of State William P. Rogers was named chairman, and former astronaut Neil Armstrong vice chairman. The 13-member panel was to report back to the President within 120 days. Its mandate: to establish the probable cause or causes of the mishap, and to develop recommendations for corrective or other action based on its findings.

Members of the presidential commission investigating the Challenger accident visited KSC and held an open hearing at the adjacent visitors center. At center is William Rogers, commission chairman, leaving KSC Headquarters with other panel members.

To support the commission, acting NASA Administrator William Graham replaced the 51-L interim panel with the STS 51-L Data and Design Analysis Task Force. The task force was then realigned to better address the commission's broad scope. Later the same month, Shuttle astronaut and Navy Adm. Richard Truly succeeded Jesse Moore as associate administrator for space flight—Moore had been named director of Johnson Space Center in early January. Truly also assumed chairmanship of the task force, and named as vice chairman James R. Thompson, deputy director for technical operations at the Princeton University Plasma Physics Laboratory and a former senior manager at Marshall Space Flight Center.

Commission members visited Kennedy for briefings, and later held a full-scale hearing at the nearby visitors center.

On June 6, exactly 120 days after it was officially created, the Rogers commission submitted its final report to President Reagan. The Challenger mishap was "an accident rooted in history," the panel concluded, dating all the way back to an inadequate joint design which neither NASA nor solid rocket motor contractor Morton Thiokol had recognized as having a serious potential for failure. Primary cause of the explosion was leakage of hot gases past the seals in the field joint between the two lower segments of the right SRB. The commission found that this failure was the result of a "faulty design unacceptably sensitive to a number of factors." These factors included the effects of temperature, reuse, processing, and the joint's reaction to dynamic loading.

The panel made nine recommendations on safely returning the Space Shuttle to flight:

I. Change the faulty solid rocket motor joint and seal design.

II. Review Shuttle management structure, placing more authority with the program manager. Increase the use of astronauts in management positions. Establish a Space Transportation System Safety Advisory Panel.

III. Review critical Shuttle parts and systems—those which, if they failed, could result in loss of the vehicle and/or life.

IV. Create a semi-independent safety, reliability and quality assurance directorate within the agency, reporting directly to the NASA administrator.

V. Improve communications.

VI. Take steps to improve landing safety.

VII. Try to implement a crew escape system usable during controlled gliding flight, and seek to increase the range of flight conditions under which an emergency runway landing could be conducted.

VIII. Establish a Shuttle flight rate consistent with available resources; avoid relying on a single launch vehicle system.

IX. Improve maintenance safeguards.

Well before the Rogers commission presented its final report to the president, NASA was already conducting its own vigorous assessment of the Shuttle program. A week after the accident, a top-to-bottom review began. "We not only went back into the solid rocket booster-solid rocket motor," said Richard Kohrs, deputy director of the National Space Transportation System (NSTS) program at Johnson Space Center in Houston, "we went back and looked at all of the elements of the Shuttle and said, 'What design improvements, what changes should we make—hardware or software—to improve margin or to improve safety?'"

Less than two months after the accident, Associate Administrator for Space Flight Truly issued a memorandum defining a comprehensive strategy and major actions required for safe resumption of Shuttle flights. Many of the requirements anticipated the Rogers commission's final report, including a complete assessment of the Shuttle decision-making process and redesign of the solid rocket booster joints.

NASA formally responded to the Rogers report on July 14, well within a presidentially stipulated 30-day deadline.

The agency was able to state that it had already initiated actions anticipating most of the nine recommendations:

I. Per Truly's March memo, the Marshall Space Flight Center was directed to form a solid rocket motor design team. Per the Rogers report, the National Research Council (NRC) established an independent oversight panel to follow the team's work.

II. In May, NASA's administrator commissioned former Apollo Program Director Gen. Sam Phillips to spearhead the review of NASA's management structure. While Phillips had broad authority to look at any and all aspects of NASA management, including relationships between the field centers and headquarters, NASA also wanted a closer look at Shuttle program management—also a concern of the Rogers commission. In June, the agency asked Shuttle veteran Capt. Robert L. Crippen to assess the NSTS program director's role, "to assure that the position has the authority commensurate with its responsibilities," and to "specify the relationship between the program organization and the field center organizations." Recommendations of the Crippen and Phillips studies would be coordinated.

III. In early March, NASA began reviewing Space Shuttle "Failure modes and Effects Analyses" (FMEAs), "Critical Items (components) Lists" (CILs), and hazard analyses. These are tools used to document what can go wrong with critical Shuttle hardware and the impact of a failure, and to insure that the criticality of the hardware is properly reflected in program documentation. Again, a separate and independent NRC panel would oversee this work.

IV. In early July, NASA formed a new top-level directorate to oversee safety operations. One task of the Safety, Reliability, Maintainability and Quality Assurance organization was to make sure NASA had enough people to ensure safety functions were adequately carried out.

V. Crippen was also charged with developing plans and policies to improve communication within NASA, and to standardize procedures for imposing and removing launch and other operational constraints.

VI. NASA established a Landing Safety Team to review and implement the commission's findings on Shuttle landing safety. The agency indicated it had initiated, some time prior to the accident, runway surface tests and landing aid requirement reviews. It had also planned previously to install advanced landing aids at the Shuttle landing facilities, and expected an improved interim brake system to be ready in 1987. Funding was already approved for an even better carbon brake system.

VII. Crew escape and launch abort studies were initiated in April, to be completed by October, with an implementation decision in December.

VIII. In March, NASA began a two-pronged study of Shuttle flight rate capability. Capabilities and constraints governing Shuttle processing flow at Kennedy were one focal point, and the impact of crew training and software delivery/certification on flight rates the other.

IX. A Maintenance Safeguard Team was created to implement the commission's recommendation and was assigned the task of developing a maintenance plan to "ensure that uniform maintenance requirements are imposed on all elements of the Space Shuttle program."

The self-examination of its policies and procedures which NASA carried out was thorough and rigorous. Every aspect of the Shuttle program was scrutinized, from the paperwork defining requirements of the Shuttle system, to the Shuttle hardware, to the pads from where that hardware is launched.

Numerous modifications were made:

- Solid rocket booster, more than 140 modifications.
- Space Shuttle main engine, about 35.
- Orbiter, more than 200.
- Kennedy Space Center's Launch Complex 39, approximately 140.

While many of the changes were made in response to the Rogers commission recommendations and the accident, others such as pad and many orbiter and engine modifications, were planned earlier. NASA simply took advantage of the stand-down in launch activity to implement them. The comprehensive reviews conducted after the accident also generated changes. For example, only about a third of the booster modifications were Challenger-driven.

A redesigned solid rocket motor, designated Qualification Motor 6, is successfully test fired April 20, 1988. Five such full-scale, full-duration tests were conducted before STS-26, each lasting about two minutes -- the approximate length of time a booster fires during actual flight.

Redesign, test and certification of the Shuttle solid rocket motor joints began in earnest less than a month after the accident. The Solid Rocket Motor (SRM) joint redesign team was headed up by Marshall Space Flight Center's John Thomas. Other Marshall officials were on the team, as were representatives from other NASA centers and government agencies, and from solid rocket motor prime contractor Morton Thiokol.

By October 1986, the team had taken the requirements for the redesigned SRM and developed the engineering improvements. Changes were made in the segment joints, and in the case-to-nozzle joints; the nozzle; propellant grain shape; ignition system; and ground support equipment. The testing program to verify the SRM redesign and certify it was extensive, including laboratory and component tests, subscale tests, full-scale simulation tests and full-scale static test firings. "All in all, the testing of the SRM redesign was a substantially greater effort than was conducted of the original design in preparation for STS-1," said Gerald Smith, SRB project manager.

Workers at KSC contributed to the redesign effort in late 1987 when they conducted a stacking exercise with the Assembly Test Article. Two pieces of the redesigned booster were assembled and disassembled to validate revised assembly and checkout procedures.

Well before Challenger, NASA had begun a program to improve the reusable Space Shuttle main engines (SSMEs). These powerful, complex engines undergo tremendous dynamic loads during a Shuttle launch, and NASA wanted to increase their safety margins and improve durability.

NASA mounted an aggressive ground test program beginning in December 1986, to certify the improvements made to the SSME. In the 12-month period that followed, 151 tests and 52,363 seconds of operation were carried out at the Stennis Space Flight Center in Mississippi—the equivalent of 100 Shuttle flights. SSMEs were fired for longer periods of time than ever before—up to 1,040 seconds, more than twice the average time during actual flight. A third test stand was constructed at Stennis to support the program, which is continuing.

From her outer thermal protective layers to the computers that guide her, the orbiter also underwent an overhaul. Again, many of the changes were already under way or anticipated before the accident—like the landing gear upgrades. Others resulted from the FMEA/CIL reviews, such as the modification to the 17-inch disconnect valves linking the external tank to the orbiter. Unintended closure of these valves would mean no propellant flow to the engines—a catastrophe. NASA added a latch mechanism to insure that the disconnect valves remain open during the thrust phase. A KSC team garnered kudos for developing the computer software to operate the delicate valve-latch system.

The addition of a crew escape system to the orbiter was recommended by the Rogers commission. Various NASA centers, the U.S. Navy, and industry escape experts participated in studies to select the best system for use during controlled gliding flight when a runway is unavailable. The system selected adds about 650 pounds (295 kilograms) to the orbiter weight. Besides installation of a pole assembly, down which each crew member would slide to clear the wing, modifications were made to allow the orbiter crew hatch to be opened. To protect a Shuttle crew member from the harsh upper atmospheric conditions at which an escape might occur, a new launch entry suit was designed. Its partial pressure, anti-exposure characteristics would protect the wearer at or below 100,000 feet (30,480 meters) altitude.

While the system can only be used in limited situations, it "provides a very real capability that we're glad to have," said astronaut Rick Hauck, who commanded the first

Shuttle to fly after Challenger. NASA also is looking into a system which can be used earlier in flight.

While it is easy to think of the recovery from the Challenger accident as one of hardware changes and testing, a massive reassessment of related paperwork was also conducted, starting with the documentation of what Shuttle hardware is supposed to do, to the Launch Commit Criteria defining countdown operating limits.

KSC conducted its own reviews of Shuttle processing paperwork, from a simple instruction sheet detailing how a single bolt should be installed, to the massive volumes detailing procedure for a launch countdown. Shuttle design centers and their contractor counterparts joined in the effort to insure that all instruction sheets used at KSC were thorough and accurate. Additional criteria were added to the Launch Commit Criteria pertaining to high-energy systems like the main engines, as well as more stringent weather criteria.

The Challenger accident marked a watershed in the history of NASA launch policy. Clearly, it was not in the nation's interest to continue toward reliance exclusively on a four-orbiter Shuttle fleet now reduced to three. The Rogers commission had warned that "such reliance on a single launch capability should be avoided in the future," and NASA agreed in its response: "NASA strongly supports a mixed fleet to satisfy launch requirements and actions to revitalize the United States expendable launch vehicle capabilities."

Both the Shuttle and expendable launch vehicles got a boost in August 1986. President Reagan announced his support for the construction of an orbiter to replace Challenger. At the same time, the White House announced that the Shuttle would no longer compete to launch commercial satellites. Given the accumulating backlog of satellites following the halt in Shuttle launches, the U.S. providers of commercial expendable launch services suddenly had a much better chance of competing for customers. The elimination of the Shuttle as a competitor paved the way for a new era in unmanned launching. Rather than supplying boosters to the government on a contractual basis, American companies would build their own boosters and launch them as well.

Commercial ELV suppliers received further encouragement the following year. In May 1987, NASA Administrator James Fletcher announced a mixed fleet concept employing both the Shuttle and unmanned launchers. The first manifest issued after the Challenger accident implemented this policy. The mixed fleet concept put redundancy back into the United States space program, reserving the Shuttle for those missions requiring its unique capabilities.

From a management perspective, NASA underwent several changes. James Fletcher, who was on hand for the inception of the Shuttle program in the 1970s, returned as administrator. Leadership of the Shuttle program was centralized at NASA Headquarters in Washington, addressing a Rogers commission concern that the NSTS program manager—previously located at Johnson Space Center in Houston—was being bypassed. Every level of Shuttle management was examined.

A new director headed up the recovery effort at Kennedy Space Center. On July 31, 1986, Center Director Richard G. Smith retired after 35 years of government service and seven years at the KSC helm. Smith won high praise from the NASA administrator and from his colleagues for his outstanding service to the space agency.

With the cause of the Challenger accident identified and NASA on track to make the changes necessary to resume Shuttle flights, Smith saw an appropriate juncture to depart. "My leaving now will allow a new director to come on board and establish himself prior to resuming our Shuttle launch schedule," Smith explained.

Succeeding Smith was Lt. Gen. Forrest S. McCartney, former commander of the Space Division, Air Force Systems Command, Los Angeles, Calif. McCartney changed some of the KSC organizational structure to match the overall agency restructuring. An additional role was assigned to the KSC deputy director, Thomas E. Utsman: director of Space Transportation System management and operations. McCartney wanted to create a single focal point within KSC Shuttle management as a liaison with NASA Headquarters program officials.

McCartney also elevated the positions of the launch and engineering directors to lessen the burden of daily administrative work. To provide stronger, independent safety oversight, McCartney established the Safety, Reliability and Quality Assurance Directorate (SR&QA), from elements formerly part of each operational directorate.

In November 1986, NASA Deputy Administrator Dale Myers announced the new management and operations structure for the National Space Transportation System. Capt. Robert Crippen, veteran of four Shuttle flights, was named deputy director of National Space Transportation System Operations, one of two top-ranking positions created to strengthen Shuttle program management. Stationed at Kennedy Space Center, Crippen had the responsibility of managing the final launch decision process, beginning with the presentation of the Flight Readiness Review approximately two weeks prior to launch, and extending through the final go/no go for launch during the countdown.

"I don't think we're doing anything totally revolutionary from the way we were doing it before," Crippen said of the revamped management structure. "The primary thing that we've done is to make sure that it is much more crisply defined, and that the roles and responsibilities of each of the individuals within it are very well defined."

Other changes were occurring at KSC. In October 1986, for the first time since the Challenger accident, an orbiter was rolled out to the pad. Atlantis remained on Pad B until mid-November to allow checkout of about $3.3 million worth of modifications designed to help shield the vehicle from harsh weather.

In addition, about 140 facility and equipment modifications were instituted at Launch Complex 39. At Pad B, the Apollo-era crew escape system design was upgraded and expanded to accommodate a Shuttle-size crew, plus other workers who might be on the service structure. Changes were also made in the Vehicle Assembly Building where the main Shuttle elements are assembled, on the mobile launcher platform upon which the assembly process takes place, and at the KSC Shuttle Landing Facility.

Despite such activities at the Center, employment dropped following the loss of Challenger. By the beginning of October 1986, the KSC work force had fallen from 16,000 employees to 13,700. Some of the decline was accomplished through normal attrition, some through layoffs resulting from the stand-down in Shuttle processing activities.

In 1987, that decline began to gradually reverse as preparations to resume Shuttle flights heated up. On Aug. 3, a sleeping beauty awoke. In High Bay 1 of the Orbiter Processing Facility (OPF), workers cheered as electrical power to the orbiter Discovery was turned on, the first major milestone in readying a Shuttle for flight. A month later, flight processing for STS-26—the first Shuttle mission after Challenger—was under way.

NASA decided to treat the STS-26 mission as if it were Discovery's first. Before the power-up could occur, many of the orbiter's major systems and components were removed and sent back to the supplier for rebuilding or modification. They were then reinstalled and checked out during flight processing. Workers also outfitted Discovery's bay to accept its primary payload: the third Tracking and Data Relay Satellite (TDRS).

The following year, early on the morning of July 4, America's Independence Day celebration got off to a historic start when the fully assembled Shuttle was rolled out of the VAB on its painstakingly slow journey to Pad B. Amidst a blaze of spotlights and waving American and NASA flags, Center Director Gen. McCartney presented STS-26 crew member Dave Hilmers with an autograph book bearing nearly every KSC employee's signature. In accepting the book Discovery's crew would carry into space, Hilmers told the crowd: "It is those of you who have written your names into this book who have made this splendid, magnificent sight that we behold tonight possible. But tonight, you haven't given this book or this Shuttle just to the five members of our crew. Indeed, you've given it to all Americans. It's the mark of a great nation ... that it can rise again from adversity, and with Discovery, rise again we shall."

Hilmers and his fellow STS-26 crew members had known of their assignment since January 1987. All were veterans of previous Shuttle flights. Crew commander was Frederick "Rick" Hauck (Capt., USN), who piloted mission STS-7 in June 1983 and commanded mission STS 51-A in November 1984. Richard "Dick" Covey (Col., USAF), who piloted mission 51-I in August 1985, would pilot STS-26. The three mission specialists were John "Mike" Lounge, who flew with Covey on 51-I; David C. Hilmers (Col., USMC), a mission specialist for 51-J, a classified Department of Defense flight in October 1985; and George "Pinky" Nelson, a mission specialist on STS 41-C in April 1984, Challenger's fourth flight, and on STS 61-C—the flight before the accident.

The long preparation period gave the crew greater freedom to participate in the modification and review process. "We've had an opportunity to get very involved in the development of the changes to Discovery," Hauck said at a preflight briefing. "We've participated in many of the management reviews. We've worked with the flight controllers and flight directors and flight designers to make sure that we've got a mission that is one we can execute and one that we're comfortable with."

There was also time to remember Challenger. "If I can speak for the whole crew," Hilmers said at the same briefing, "we believe the Challenger accident was a great tragedy to our nation. It was a great tragedy to each of us personally. On the one hand, we can't forget what happened, because if we do, then we're prone to make the same types of mistakes that we made in the past.

"But on the other hand, we can't dwell on what happened in the past, or we'll never look forward to the future. In the same way, when we're on the launch pad, we'll be thinking about them. We'll have them in our minds, but I think our thoughts will be looking more forward and upward than they are into the past."

The STS-26 crew arrives at the KSC Shuttle Landing Facility in April 1988 for return-to-flight activities. From left to right are Mission Specialists Dave Hilmers and Mike Lounge, Commander Rick Hauck, Mission Specialist George Nelson and Pilot Dick Covey. All had prior Shuttle flight experience and were an enthusiastic, articulate crew.

The targeted launch day was sometime in late September. Because of the two-year launch hiatus, and the extensive facility and hardware modifications, the prelaunch schedule at KSC was a grueling one. Three major prelaunch tests at the pad were conducted: the Wet Countdown Demonstration Test, which demonstrated the performance of the main propulsion loading system; the spectacular Flight Readiness Firing of the three main engines; and the Countdown Demonstration Test, a dress rehearsal for launch in which the flight crew participated.

With the three key tests accomplished and test data coming out clean, KSC workers continued to prepare for launch. Associate Administrator for Space Flight Truly flew down from NASA Headquarters in Washington to chair the Flight Readiness Review Sept. 13-14. During this top-level meeting at KSC's Operations and Checkout Building, NASA managers certified—and backed up the certification with data—that all was in readiness to proceed with a launch.

When NASA announced that Discovery would be launched sometime within a three-hour window the morning of Sept. 29, the long and sometimes lonely struggle to recover from the aftermath of the Challenger accident approached a happy resolution. On Sept. 26, an exuberant flight crew zoomed into KSC's Shuttle Landing Facility in their T-38 aircraft and pronounced that they were ready to go. Commander Hauck succinctly stated: "The bird is ready and we're ready. We're excited and we cannot wait to fly."

The STS-26 launch countdown clock was already ticking. Hotel rooms in towns near the Center filled up rapidly, and trailer homes sprouted up along the excellent vantage points on the Indian River shoreline. An estimated 250,000 people came to watch the resumption of U.S. manned space flight. The media and contractor public relations crowd—about 2,500—was the second-largest in history, topped only by the first Shuttle flight in April 1981.

At 11:37 a.m. EDT on Sept. 29, 1988, the Space Shuttle Discovery and her five-member crew blast off from Pad 39B, signifying a new beginning for America's manned space flight program.

The launch window for STS-26 extended from 9:59 a.m. to 12:29 p.m. In the preceding hours, KSC and its environs were astir with constant activity as the countdown proceeded. The KSC News Center remained open all night. Media representatives afraid to leave the center for fear of missing the launch tried to sleep in their cars. Helicopters buzzed in and out of the landing pad near the Vehicle Assembly Building, and fixed wing aircraft could be heard periodically passing over. Workers arriving in the dusky predawn hours passed expectant onlookers gazing toward Discovery, regally poised in the distance on the brightly lit Launch Pad B.

The morning of launch, the flight crew was awakened 25 minutes earlier than in the past to allow time to don their new—and bulky—launch entry suits. Unusually light upper winds delayed liftoff, but at 11:37 a.m., Discovery roared off the pad and put the American manned space program firmly back on track.

Upon their return Oct. 3, 1988, Discovery's crew was greeted by Vice President George Bush and an ecstatic crowd several hundred thousand strong.

Discovery's mission was a conservative one, lasting only four days. "I think we all would have liked to have spent more days in space," Hauck commented before the mission, "but it's appropriate that after being thrown from the horse, we trot before we gallop." The TDRS spacecraft was deployed about six hours into the flight. The remainder of the mission was devoted to the 11 middeck experiments, taking remote photos of the Earth and determining how long it took to don the launch entry suits in the low-gravity environment. The crew also held a memorial service to honor the Challenger crew.

Only minor problems—among them a faulty television camera and a malfunctioning cooling system—marred the flight. After logging nearly 65 orbits around the Earth, Discovery and her crew touched down at Edwards AFB in a picture-perfect landing at 12:37 p.m., Oct. 3. "That's a great end to a new beginning," commented astronaut Blaine Hammond from Houston. A tumultuous welcome from more than 400,000 well-wishers, including Vice President George Bush, awaited a jubilant crew.

Looking as good or better than any previously flown orbiter, Discovery returned to KSC Oct. 8 atop the 747 Shuttle Carrier Aircraft. More good news came from the redesigned solid rocket booster's performance: No signs of gas leakage in the joint areas were detected.

"As a crew, we were tickled pink at Discovery's performance," Pilot Dick Covey said after the mission. "That's a real credit to the crews at KSC who got it ready for flight." On Oct. 25, a grateful Discovery crew returned to the center to say thanks in person. They were regaled en route with an old-fashioned parade, a tradition dormant since the space program's earlier days.

Even as the celebration of STS-26 was under way, KSC was busy preparing for another launch. STS-27 had none of the fanfare which surrounded Discovery's historic flight. An all-military crew was named to carry out this classified mission for the Department of Defense. Robert L. Gibson (Cmdr., USN) was mission commander; Guy S.

Gardner (Lt. Col., USAF), pilot; Richard M. Mullane (Col., USAF), William M. Shepherd (Cmdr., USN), and Jerry L. Ross (Lt. Col., USAF), mission specialists.

Atlantis lifted off at 9:30 a.m on Dec. 2 into a clear blue sky. Because it was a classified mission, very little detail was provided about the flight. The orbiter touched down at 6:36 p.m., Dec. 6, after four days and nine hours in space.

The diverse capability of the Shuttle program was once again demonstrated in May the following year, when an orbiter carried aloft for the first time a planetary explorer spacecraft. The successful deployment of the Magellan Venus radar mapper by Atlantis on STS-30 marked the resumption of U.S. planetary exploration. Subsequent Shuttle missions will carry additional scientific spacecraft as the United States renews its exploration of the solar system.

By the end of August, all three orbiters had flown at least once. The return to flight was completed Aug. 21 when Columbia, the oldest orbiter in the Shuttle fleet, returned to KSC's Shuttle Landing Facility from Edwards Air Force Base. Columbia's launch Aug. 8 on a five-day mission for the Department of Defense marked the first time it had flown since January 1986.

Center Director Forrest McCartney praised the KSC employees who had worked long and hard to safely return the Shuttle to space. "The success of your effort has rekindled the spirit of a grateful nation and renewed the desire of Americans everywhere to continue to push back the frontiers of space. As we at Kennedy move forward into a new era of exploration, each milestone, each task accomplished will strengthen and reinvigorate the unwavering pride and commitment that is our heritage," he wrote in an open letter to KSC workers.

As the final decade of the twentieth century unfolds, the Shuttle will continue to play a vital role in the U.S. space program. In 1989, President George Bush proclaimed the long-term goal of returning Americans to the moon and after that, moving onward to Mars. NASA will play a major role in creating the blueprint for the two missions. A giant first step in the return journey could occur when the Space Station Freedom becomes fully manned and operational. The Shuttle will be the primary vehicle for carrying parts of the station into space, where they will be assembled. In 1991, the existing orbiter fleet was joined by Endeavour, Challenger's replacement. When Endeavour flies for the first time in 1992, NASA will have a full complement of working Space Shuttles leading the way toward turning present-day dreams into a future reality.

Chapter 16

CONTRACTORS at KSC

The national space program is a joint undertaking of the federal government, private industry and the educational community. Companies and institutions providing products and services to NASA account for the bulk of the agency's annual budget. At their peak in 1966, the nation's manned space flight programs alone employed more than 300,000 persons, most of them in industrial plants and universities from coast to coast.

Kennedy Space Center staffing illustrates the melding of the country's government and private sectors into a single integrated team to execute launch missions. Of the work force of more than 17,000 engaged in KSC activities, less than 3,000 are federal employees, while the remainder work for resident contractors.

It is NASA policy to award contracts competitively. Contracts for flight hardware—launch vehicles and spacecraft—are let by NASA development centers, which monitor the contractors' efforts throughout design, development, testing and evaluation phases of the various flight articles. The Johnson Space Center, for example, monitors the performance of Rockwell International, the firm chosen by NASA to design and build the Space Shuttle orbiter. The Kennedy Space Center usually employs the same contractors, at least initially, to assist in checking out, assembling, testing and launching the flight hardware once it arrives at the spaceport.

KSC also awards contracts for support services. These contractors do not build a product, but they provide specialized services such as technical and administrative communications, instrumentation, facilities engineering, and housekeeping. Unlike hardware or element contractors (vehicle stages, guidance systems, spacecraft, etc.), which are usually chosen for the life of a particular program, support contractors are selected for definite time periods, of three to 10 years in most cases.

KSC's Industry Assistance Office is the primary interface with new companies seeking to do business at KSC. The office receives and opens bids, arranges demonstrations by companies at KSC, sponsors off-site business procurement conferences and maintains a computerized bidders' source list. Each year KSC sponsors a procurement briefing to inform industry of upcoming contract opportunities at the Center, an initiative begun at KSC in 1982 and since adopted by other NASA Centers.

Many support function contracts, such as data processing, mail delivery, library and custodial services, are reserved for small firms. Both NASA and NASA contractors at the Center seek to assure small business and minority-owned companies a role in the nation's space program. The Industry Assistance Office participates in and sponsors industry conferences and briefings at both the state and local level. KSC is the only NASA Center which honors its small business contractors with annual awards.

Contractors provide a wealth of services at KSC and constitute the bulk of the Center's work force.

Reproduction workers stand behind their product, above. Some members of a launch team wave flags after another successful liftoff, below.

Ensuring equal opportunity for all KSC employees, both contractor and civil service, is the charter of the Equal Opportunity Program Office. This office is responsible for developing and administering NASA's equal opportunity policy at the Center. A variety of services are available, including training, counseling, referral or intervention. The office is staffed with a director, a federal women's program manager, an Hispanic employment program manager, and an affirmative action/complaints coordinator. The federal women's program manager and Hispanic employment program manager are each responsible for working groups whose members represent every major directorate at KSC. Through these working groups, KSC management is kept informed of the concerns of women and Hispanic employees.

Sometimes misidentified as a member of the contractor work force is the company which operates the KSC visitors center, Spaceport USA. TW Recreational Services Inc. is actually a concessionaire. While a visit to the popular tourist attraction is free, there is a small fee for taking a bus tour of KSC or Cape Canaveral, and for one of the film experiences offered. The concessionaire makes a profit and, at the same time, invests funds in construction of additional facilities and exhibits.

Contractor organizations have been assimilated into the local community just as they have become integral elements of Kennedy Space Center. They participate in community affairs, contribute to charities and take active roles in scientific and technical societies which have local chapters and which are attracted to the area's space environment for regional and national meetings.

To promote team spirit and enhance morale among government and contractor personnel, the activities of the Kennedy Athletic, Recreational and Social Organization, or KARS, sponsored by the NASA-KSC Exchange Council, are open to all Center employees. This includes bowling and softball leagues, golf matches, skeet shooting, boating and many other group activities.

KARS has also developed two recreational areas, which offer facilities for such outdoor pastimes as camping, fishing, racquetball, tennis, and softball. KARS Park I is located in the southern area of the Center along the Banana River, and KARS Park II is located north of it along State Road 3, west of the KSC Industrial Area. Exchange Council projects are operated on non-appropriated funds derived primarily through receipts from cafeteria and vending machine operations.

The following list is representative of the types of activities contractors and subcontractors perform at KSC:

All World Travel - on-site travel service

Atlantic Technical Services - mail and distribution services

BAMSI Inc. - operations and maintenance services

Bechtel National Inc. - architectural and engineering services

The Bionetics Corp., Biomedical and Environmental Laboratories, and Shuttle Calibration Laboratories - laboratory and environmental support services; standards and calibration

Boeing Technical Operations Inc., Aerospace Operations - ground system operations, automated management, Spacelab experiment integration support

Computer Sciences Corp., TICS Project - Shuttle inventory management, communications and instrumentation support, office automation

Ebon Research Systems - reliability and quality assurance services for safety engineering

EG&G Florida Inc. - base operations

General Dynamics Space Systems Div. - Atlas-Centaur launch operations, support and engineering activities

GTE Automatic - administrative telephone service

Honeywell Federal Systems Inc. - computer operations and maintenance

IBM Corp. - systems engineering and software for Launch Processing System

Lockheed Space Operations Co. - Shuttle processing

McDonnell Douglas Space Systems Co. (Cape side and KSC) - payload processing and ground operations

Martin Marietta Manned Space Systems - manufacture and checkout of Shuttle external tank

New World Services Inc. - library services

RCA International Service Co. - communications and instrumentation support services

Reynolds, Smith and Hills - architectural and engineering services

Rockwell International Corp., Rocketdyne Div. - manufacture and checkout of Space Shuttle main engines

Rockwell International Corp., Space Transportation Systems Div. - manufacture of and logistics support for orbiter

Specialty Maintenance and Construction Inc. - design and construction services

Sun Coast Services Inc. - food service concession agreement

TGS Technology Inc. - photographic services

Teledyne Brown Engineering - payload integration operations

TRW Defense Systems Group, TRW Space and Technology Group - safety, reliability and quality engineering, and defense and space systems

Unified Services Inc. - custodial support

United Technologies Corp., Chemical Systems Div., and USBI - separation motors on Shuttle solid rocket boosters; design, assembly and refurbishment of booster non-motor segments

Wiltech Corp. of Florida Inc. - component refurbishment and chemical analysis services

Washing the parachutes that lower the Shuttle's reusable solid rocket boosters into the ocean is all in a day's work for the KSC contractor employees.

chapter 17

TECHNOLOGY TWICE USED

Doctors in Houston monitor the heartbeats of astronauts in space—and the new techniques are adapted for cardiac patients on Earth.

A new understanding of how metals crack prevents rocket failures—and engineers apply the knowledge to bridges and pipelines and buildings.

New fire-resistant coatings and fabrics shield spacecraft and aircraft—and now protect railroad cars and firemen and homes.

American industry develops bold new concepts for space equipment—and uses the new capabilities for better products.

Our ventures beyond our own planet have called forth the best efforts of America's imaginative scientists, engineers, and managers. Meeting the challenge of exploring and using space for the benefit of humanity has expanded knowledge and skills in virtually every field of science and technology. Creative people in all walks of life have recognized the value of this new knowledge and its new technologies and are utilizing them to make life better for everyone.

Technologies developed for specific purposes often can be applied in other areas. These applications are known as indirect benefits, or "spinoffs." For instance, NASA's requirement for small, lightweight, dependable guidance and communications systems for spacecraft brought about electronic miniaturization and a revolution in computer technology. An excellent example of this spinoff is the pocket calculator. Technologies developed through NASA's far-ranging aerospace programs have found applications in thousands of areas—the list extends to catalog length. Collectively, they add up to consequential gains in personal convenience, human welfare, industrial efficiency and economic value.

At KSC, the Technology Utilization Office within the Advanced Projects, Technology and Commercialization Office is involved in making space technology available to private industry. This office is charged with the task of informing people and organizations about currently available aerospace technology, and helping them put this information to work. It tries to accomplish this in two ways: by helping people from industry and educational institutions find available technology that fits their needs; and by identifying and publicizing new technologies developed at KSC and making them available to potential users.

The Center's patent counsel works with KSC employees who have developed equipment or ideas that may be patentable. Inventions patented by NASA can readily be licensed, and NASA encourages their commercial use. The KSC Technology Utilization Officer reviews proposed aerospace transfer projects to determine whether they comply with established criteria such as technical feasibility and cost.

The Shuttle Launch Processing System presented an opportunity for technology transfer on a huge scale. This highly automated system was designed and developed at KSC for Space Shuttle checkout and launch operations. KSC researchers developed a plan to further apply the complex concepts behind it. These pioneering approaches can be utilized in designing control systems for many functions, including such sensitive operations as nuclear power plants. One day the creative engineering work performed at KSC may be the basis for a system that automatically checks out malfunctions in atomic reactors, or provides an automatic "shutdown" order in time to prevent damage during an emergency.

One of the major components developed for the Launch Processing System is the Common Data Buffer, which can serve as the interface and communications medium in non-aerospace computer complexes. It can be used with any computer and for processing in any computer language, making it widely applicable throughout the industry.

On a smaller scale, another innovative transfer of space technology involves firefighting equipment designed for use in the event of a Shuttle orbiter crash landing. Developed by KSC and Boeing engineers, the fire extinguisher has a hard pointed tip capable of piercing the orbiter's outer layers. Fire-extinguishing chemicals can then be injected inside the spacecraft. After obtaining a license from NASA for commercial use of the technology, a Massachusetts-based company introduced a product used primarily by airport firefighters. The 82-inch (208 centimeter) nozzle device, weighing about 30 pounds (13.6 kilograms), has a tip of hardened steel. It has proven most useful in combating aircraft fires. Fire-dampening chemicals or water are discharged through the nozzle into an aircraft's passenger cabin, cargo compartments, accessory bays or ducts.

Space technology is finding its way into the home as well, thanks to collaborative efforts between KSC, private industry, state and federal agencies, and utilities. About 70 percent of our orbiting satellites incorporate heat pipes, which cool critical electronic components in the spacecraft. Skylab and the Hubble Space Telescope also have utilized this energy-saving technology. In its commercial application, the pipe is used in conjunction with the air conditioning system to cool and dehumidify air more efficiently. A Florida-based entrepreneur has developed a system that can be installed in the home, with anticipated energy savings of 15 to 20 percent.

The state of Florida is working with KSC to reap the benefits of space technology. A property appraiser adapted aerial infrared mapping technology so it could be used to inventory citrus trees as a basis for determining citrus grove valuations. KSC, working with the University of Florida's Citrus Research and Education Center, came up with a dual video system to interpret the photos. Aerial photos of citrus groves taken over a period of time are compared by the video system, revealing changes from one year to the next.

This example of innovative technology transfer resulted in more accurate property valuations at a lower cost. Citrus growers also can use the data to determine areas where problems may exist. The Florida State Department of Revenue is interested in a more automated system which could be used in areas with very large citrus acreages. KSC has issued a contract to the University of Florida Citrus Research and Education Center to develop such a system.

The Technology Utilization Office has helped develop applications for weather satellite data in a joint project with the National Weather Service and the University of Florida. Hourly surface temperature distribution data from the satellite can be analyzed by a computer, in conjunction with ground measurements of meteorological data, to produce a forecast of temperature distribution throughout the remainder of a cold night. Both the observed temperature, which can be measured to an accuracy of about 2 degrees Fahrenheit (16.7 degrees Celsius) by the satellite, and the predicted temperature are valuable tools the forecaster uses in developing his frost-warning forecast.

The list of NASA contributions to the health care field continues to grow. A system which has added to the store of human knowledge of the Earth is also enhancing our knowledge about our own bodies. KSC is working with medical researchers to apply the image processing techniques used in the Landsat remote sensing program to a new diagnostic tool called Magnetic Resonance Imaging (MRI).

Still largely an experimental technique, MRI uses a magnetic field and radio waves to create body images. Unlike potentially dangerous X-rays, MRI can penetrate bone. In addition, it is a non-invasive technique which is capable of producing large quantities of anatomical and physiological data.

The disadvantage of MRI, radiologists found, was that much of this data was redundant. It also meant there was more to look at, taking up time and increasing the diagnostician's workload. By applying Landsat processing technology, researchers are streamlining the MRI imagery to come up with a composite picture. Landsat, the NASA-developed Earth remote sensing satellite, takes multiple pictures of the Earth. The huge quantities of raw data which it sends down to Earth are analyzed by a computer program. This program cleans up the data—sharpening contrasts in images and eliminating confusing detail. The result is false color Earth imagery with each feature defined by using different colors.

When the MRI data was plugged into the Landsat processing computer program, the results were equally useful. One radiologist said it was as though a slice of the human body had simply been lifted out for scrutiny. Researchers are striving to take this amazing example of technology transfer even further. They are creating "theme maps" of the human body. The data and computer software can be programmed to look for a particular signature in the MRI imagery, such as a blood clot in the brain or a tumor.

The researchers are now working to convert the Landsat computer program to make it compatible with the computer used in the MRI system. This would allow expansion of the technique to all locations where MRI is employed. As one radiologist put it: "Even these first crude experiments show that the potential is very great. Satellite imagery has opened a new window into the human body."

Besides answering direct queries to KSC, the Advanced Projects, Technology and Commercialization Office also supports STAC—the Southern Technology Applications Center. STAC helps Florida-based and other industries in the southern United States find solutions to their problems. STAC provides ready access to more than 10 million articles stored in computer memories. Besides NASA, Florida universities and the Florida Department of Commerce also support STAC.

In addition to direct support from such sources as the Technology Utilization Office at KSC, would-be entrepreneurs, researchers and other interested individuals and companies can turn to two NASA publications for information. NASA Tech Briefs, published monthly, detail technology advances being made at KSC and other NASA Centers. An annual publication, Spinoff, reports on the products resulting from technology transfer.

KSC's Technology Utilization program helps businesses avoid the costly process of "reinventing the wheel" when the technology they need is already available. This returns the taxpayer's money in the form of immediate benefits.

Thousands of applications with a genesis in aerospace technology are already benefiting the public. Some of these are genies of convenience, bearing the luxuries of comfort and speed. But other, more important spinoffs convey necessities like food and warmth, which are crucial to human survival. Countless new spinoffs are expected in the years ahead as the Space Shuttle's broader capabilities begin to yield proportionately greater benefits.

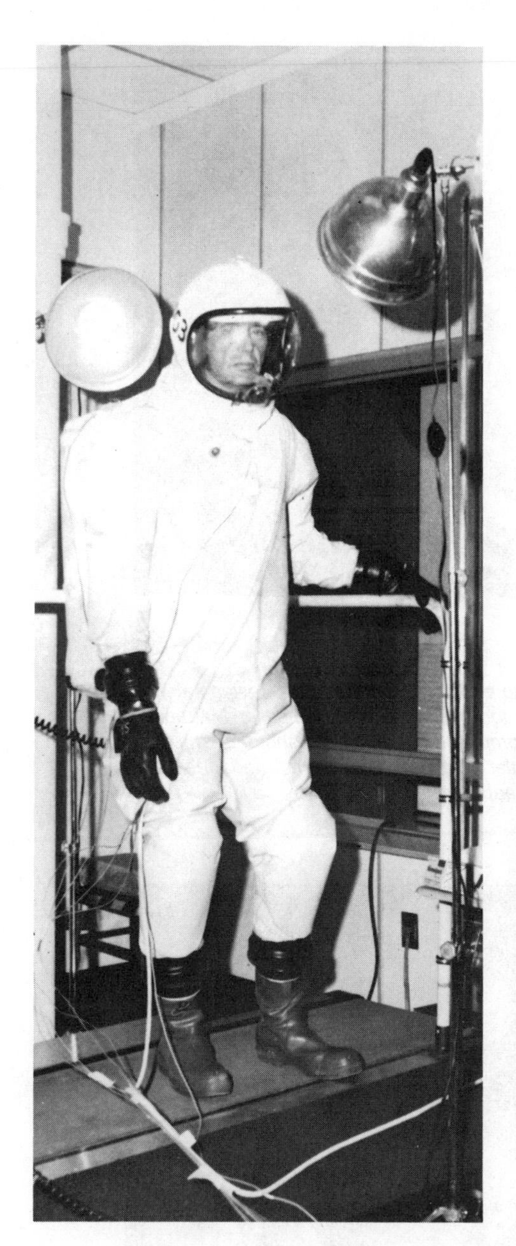

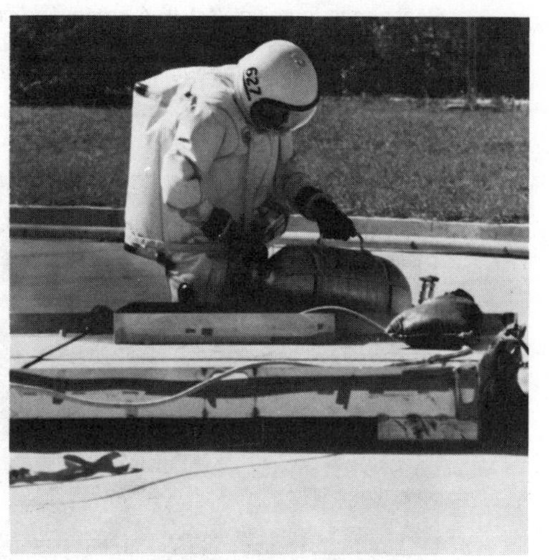

Protective equipment, such as the Self-Contained Atmosphere Protective Ensemble suit worn here during training exercises, is finding uses in industry.

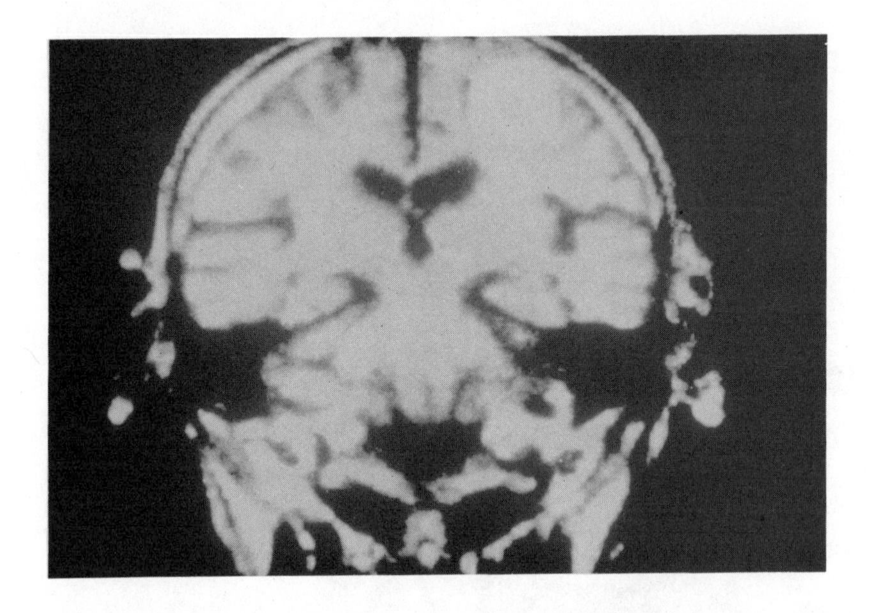

The computerized image processing used to enhance Earth remote sensing satellite photos is being applied to a new diagnostic tool in the medical field called Magnetic Resonance Imaging (MRI). A processed composite image, upper photo, of an MRI head scan shows a brain tumor (white circular area near upper right), and further enhancement defines the tumor even more clearly, lower photo.

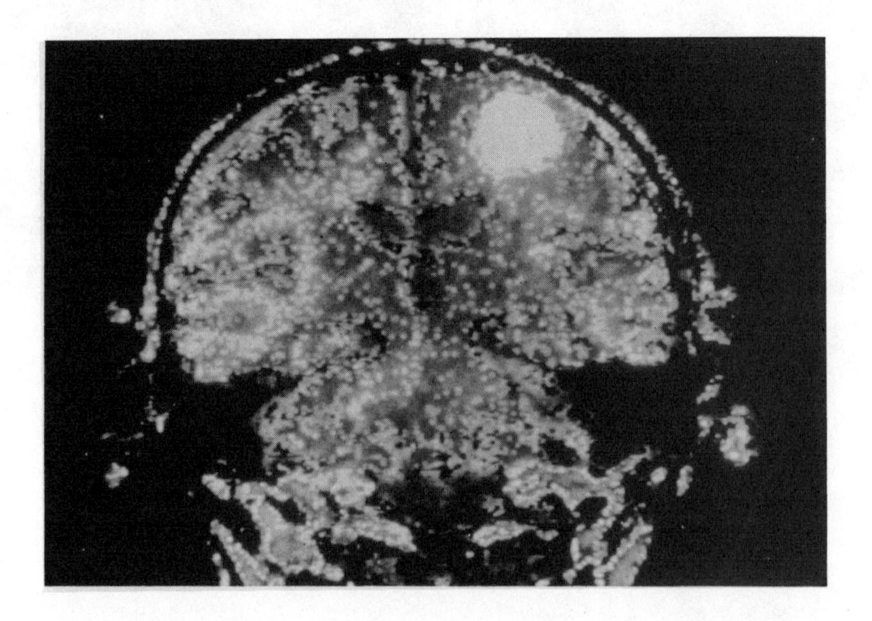

chapter 18

COMMUNITY, PUBLIC and PRESS

The growing space program brought profound changes to the Florida coastal communities surrounding the Kennedy Space Center. Brevard County, in which the Space Center is located, became Florida's 25th county in March 1844. However, the changes the county experienced between 1963 and 1969 occurred faster and touched more people than those during any comparable period in the previous 119 years.

Prior to the space boom, Brevard's economy was largely based on citrus production. Of 839,404 acres (335,762 hectares) in the county, more than 20,000 acres (8,000 hectares) were cultivated to produced the famed Indian River oranges and grapefruit.

In 1950, Brevard's population, which depended mainly upon this agricultural resource, was 23,700. When new government programs began developing missiles for defense, the capability of the Eastern Test Range kept pace and launch complexes and industrial facilities at Cape Canaveral Air Force Station were constructed. Thousands of government and contractor employees streamed in to operate the Range and conduct launches. This explosive growth raised Brevard's population to 91,900 by 1958. Yet sizable as the expansion prior to 1960 was, the decision to undertake the Apollo program and the choice of north Merritt Island as the launch base caused a much heavier impact. Population soared to 247,500 in the next 10 years.

The state of Florida joined with NASA and the Air Force to help the community solve the problems that accompanied rapid growth. A Joint Community Impact Coordination Committee was formed. It provided information to the community regarding the phased buildup of the work force. In turn, the community advised the committee of anticipated problems, the need for federal assistance and plans to solve the problems. The committee functioned as a catalyst in relation to federal agencies which could provide help in constructing and expanding roads and bridges, water supplies, hospitals and schools.

The Federal Housing Administration encouraged the construction of homes and apartments by underwriting loans. Federal aid also assisted in building new facilities at the Melbourne Regional Airport and the Space Center Executive Airport, south of Titusville, near KSC.

NASA and the Air Force contributed directly to solving some problems. When traffic bottlenecked on the two-lane, low-level bridges crossing the rivers between Merritt Island and its neighboring towns, the county appealed to Vice President Hubert H. Humphrey. At his urging, NASA and the Air Force each contributed $2 million to help finance high-rise four-lane bridges.

These collective efforts continued into the spring of 1965. Subcommittees representing municipal governments, civic groups and business interests pursued specific projects. An East Central Florida Regional Planning Council was organized. Jointly supported by the seven contiguous counties, it took an active part in community planning. By 1965, the area was well on the way to meeting its needs, and the Impact Coordination Committee was disbanded.

Brevard's period of accelerated growth did not last forever. Although KSC manpower grew, employment at the Air Force Eastern Test Range began to decline. In 1967, 16,710 personnel in the county were employed by the Range. By 1971, employment was down to 14,881.

Employment at KSC peaked in 1968, with 26,500 people working on the Apollo program. But by July 1970, national priorities had been realigned and, in the harsh light of reduced funding, KSC's work force was pared to 15,000. The unemployment rate in the county soared because there were no local industries to absorb the engineers and technicians now without work.

Brevard County followed three avenues to stabilize and diversify the economic base: small industries, not related to space technology, were encouraged to locate in the area; realtors actively solicited retired people to purchase homes; and the county's promotional organizations pushed for more tourists.

Midway through 1971, the economic situation began to improve. Retirees found the housing market attractive. Condominium construction increased, and retail sales rose.

Following Skylab, employment at the Center dropped to around 8,000, where it remained until Shuttle activities edged it up to a 1981 complement of about 12,000, and 15,000 by 1985. In the late 1980s, with a healthy economy and the resurgence in unmanned launches and the build-up of the Space Shuttle program, the Brevard County population continued its upswing. The Titusville-based Space Coast Development Commission projected the county would have nearly half a million inhabitants by the year 2000. At KSC, the employment level rebounded from its post-Challenger slump of 13,700 to about 17,000 in the late 1980s. The extent of KSC's contribution to the Florida economy is substantial: more than $1 billion in fiscal year 1988 alone through jobs and contracts.

* * * * * *

During its history, KSC has been host to scores of distinguished visitors: presidents of the United States, heads of foreign nations, senators, representatives, ambassadors, business leaders, prominent educators, and foreign delegations, including cosmonauts. Official guests have toured the facilities, attended launches and participated in special events held at the Center.

The official visitors, however, have been far outnumbered by the general public. Kennedy Space Center has become Florida's fourth most popular attraction for visitors to the state. More than 42 million people have toured the spaceport, eager to see at closer range the place where America's space program began and continues to flourish. The number of visitors has steadily increased, and today averages about three million annually.

Official visitors, such as Congressman Bill Nelson, pictured here with his son during the launch of an Indonesian communications satellite aboard an expendable launch vehicle, are far outnumbered by the millions of members of the general public who regularly attend tours and launches.

In anticipation of public interest in NASA's activities, the congressional act which created the agency in 1958 charged it with the responsibility to widely disseminate the results of its projects. In 1963, U.S. congressional Representative Olin Teague of Texas, chairman of the House Manned Space Flight Subcommittee, raised the question of public access to the Center.

NASA Administrator James Webb pointed out that NASA had to comply with the policies of the Department of Defense, which controlled access to Cape Canaveral. However, NASA would consider allowing tours of its installation after construction was completed.

However, that same year the Air Force decided to open Cape Canaveral to the public. Motorists could drive through the installation during a three-hour period each Sunday. As they entered the gate, they received booklets which traced the tour route and provided information on the facilities. Visitors were not permitted to stop, but they could take photographs from their moving cars.

Facilities on Merritt Island were sufficiently advanced by November 1964 to permit the same type of Sunday drive-through tour. Visitors could tour either or both installations, using the NASA Causeway which linked Cape Canaveral with U.S. Highway 1 on the mainland. The public's response was enthusiastic.

NASA eventually extended the drive-through policy to include national holidays and Saturdays, and increased the access hours from three to six. The Center set up a modest collection of model rockets and pictures in a warehouse for visitors to view. The attendance at this makeshift museum grew rapidly as the Gemini program began.

Early in 1965, the National Park Service recommended a means of satisfying the public's curiosity about space. A 100-page report predicted a steady buildup in attendance. These projections were based on the assumption that a suitable visitors center would be constructed, that escorted bus tours would be conducted daily and that NASA would develop a program to enhance understanding of space exploration and the activities conducted at KSC.

Based on the study, an architectural firm drew designs for the visitors facility. NASA selected a site on its own property, one mile (1.6 kilometers) west of the Industrial Area.

Guidelines were established for the bus tours. Visitors would be permitted to drive to the new center, where they could view exhibits and films and attend aerospace lecture demonstrations at no cost. If they took the bus tour, they would pay a modest fee to cover its cost. The free drive-through tour would continue.

NASA contracted with a commercial firm to operate the visitors center and the bus tour. Temporary structures were erected on NASA-owned property bordering the Indian River on the mainland. A small exhibit building, bus terminal, and parking lot were readied, with adjacent souvenir stands, snack bars, restrooms and business offices. The General Services Administration rented 10 overhauled buses—each of which had traveled two million (3.2 million kilometers) miles—to start the tours. These were leased in turn to the contractor, which employed escort-drivers. NASA laid out the tour routes, coordinating the portions on Cape Canaveral with the Air Force.

The bus tours began in July 1966 with a single 55-mile (89-kilometer) trip that lasted more than two hours. On the Cape side, patrons stopped at the Mission Control Center used for Mercury launches and at the Air Force Space Museum, where they could photograph rocket displays. At Complex 39 on KSC, they stopped at a Saturn V launch pad and entered the transfer aisle of the Vehicle Assembly Building.

Visitors wishing to take a guided bus tour today may elect to go on the Cape-side trip, which stops at many historic sites, including the early Redstone and Mercury launch pads where America's space program began. A second option is the KSC tour. The trip inside the Vehicle Assembly Building had to be discontinued due to Shuttle operational requirements, but most buses can still drive past the Launch Complex 39 pads, and visitors can photograph the Apollo launch sites that have been converted to accommodate the Shuttle.

Next to the vast Vehicle Assembly Building, visitors can view another giant, the Saturn V rocket that carried men to the moon. Displayed horizontally, this exhibit was part of "Third Century America," the U.S. Bicentennial Exposition on Science and Technology, held at KSC during the summer of 1976. Also included on the KSC tour are visits to a huge simulated Apollo firing room in the Engineering Development Laboratory, where the launch of Apollo 11 and the first lunar landing are re-created—using the actual equipment and the recorded voices of the astronauts and launch team.

Exhibits at Spaceport USA, such as the "Rocket Garden" with its full-size launch vehicles and impressive items of space hardware, have expanded.

Popular with visitors who tour the Center is this giant Saturn V rocket, an enduring symbol of the Apollo lunar landing program. The booster is not a model, but actual flight hardware left unused when the program ended.

As the popularity of the tour program grew, KSC pressed forward to obtain larger, permanent facilities. Escalating construction costs made it impossible to build the original design. So it was decided to prepare the chosen site west of the Industrial Area, install utilities, and erect interim prefabricated structures. These buildings opened in August 1967. Two auditoriums, each seating 250, as well as a snack bar, souvenir sales area, and ticket counter, were provided.

Since renamed Spaceport USA, the visitors center has been expanded many times to better serve the increasing number of guests and to provide new displays. Additions include the Hall of History, where space equipment and technology are displayed, and the Carousel Cafeteria. A huge new theater complex, the Galaxy Center, features two theaters and three exhibit areas. "The Boy from Mars," a recent addition to the film collection, offers a vision of the future, when colonization of space and Mars has occurred and the first child born on the red planet learns about the Earth from which his ancestors came.

There is a charge for one theater, the IMAX. The 70 millimeter IMAX projection system uses a curved screen 70 feet (21.3 meters) wide and 50 feet (15.2 meters) high, supported by a six-track stereo sound system, to show imagery taken largely in space. The viewer experiences, as realistically as possible, the thrilling sights and sounds of space flight. IMAX movies are only available at a few places throughout the world.

Life on a futuristic space station is revealed in "Satellites and You," a state-of-the-art multimedia presentation recently added at Spaceport USA. Visitors can also learn more about the plants and wildlife indigenous to KSC's vast wildlife refuge from an exhibit sponsored by the U.S. Fish and Wildlife Service.

Spaceport USA is open every day of the year except Christmas, and for limited periods during Shuttle launch operations.

* * * * * *

Educational activities play a major role in NASA's program of disseminating information to the public. The Kennedy Space Center's education staff serves a region embracing Florida, Georgia, Puerto Rico, and the Virgin Islands, providing educational resources to professional educators, students of all grade levels, civic groups and various organizations such as PTAs and Scouts.

The educational community is kept up-to-date on NASA research and development programs through conferences, technical briefings, educational television and teacher programs that vary from one-day seminars to two-week participation workshops. "Teacher kits," containing a wide assortment of comprehensive literature, are mailed to instructors on request.

Aerospace science and technology talks and demonstrations are provided at a "hands-on" activities center at Spaceport USA, with material suitable for primary and secondary level students. Educational support is provided to planetariums, science centers, science fairs and youth programs. Certificates of outstanding achievement are presented yearly at Florida and Georgia state science and engineering fairs to award winners selected by NASA personnel from KSC. In addition, one outstanding science fair winner of suitable age is awarded a summer job at KSC each year.

Summer jobs at KSC are also awarded each year to a group of academically talented students who compete with their peers to win the appointments. These Summer High School Apprenticeship Research Program (SHARP) students undergo an intensive work/learning experience in the area of their particular career interests, under the sponsorship and direction of a working KSC scientist or engineer. This program, in effect since the 1979/80 school year, helps determine career choices for some of the brightest local area students by exposing them to the "real world" of high technology and science.

The education staff also is responsible for answering the more than 100,000 queries from the general public which KSC receives each year. Each letter receives individual attention. Requests from students can usually be met with a kit of more than 20 items, including Shuttle mission reports, space program history, material about astronaut selection and training, and brochures on all phases of operations from launch preparation to liftoff and landing.

Larger kits specially tailored to either teachers or libraries are also sent out, and can be customized to suit individual requests. Altogether, an average of 10,000 responses to queries are sent out monthly during the school year, with the numbers typically increasing in the periods immediately before, during and after a Shuttle launch.

The Spacemobile Program is another activity that demonstrates NASA's commitment to aerospace education for the schools and the general public. KSC sponsors two vans, which cover Florida and Georgia. Other NASA Centers cover different parts of the country. These units, manned by experienced teachers, come equipped with models, dynamic exhibits, films, slides and other visual aids. In effect, the aerospace specialists operating these vans take the classroom to the student, working with teachers and school science departments throughout the region to bring space technology directly to the students. In a typical year, the two KSC Spacemobile lecturers conduct some 1,200 demonstrations involving more than 250,000 students.

The KSC education staff also works with NASA Headquarters and other NASA Centers in bringing various groups of educators to KSC to tour the Center, receive briefings, and see Space Shuttle launches.

* * * * * *

The "open door" policy which admits the press to cover NASA launches was instituted in 1958, completely reversing the government's position on military launches during the 1950s. In those days the military services secretly conducted test launches of ballistic missiles from Cape Canaveral. During early operations, even the shape of the rocket standing on the pad was concealed. Except for the military, government, and contractor employees involved in the programs, security precautions demanded that no one know the date and time of the launch, what would be tested, and what happened after launch.

Reporters avidly grasped at rumors, and watched motel registers for the telltale names of managers and engineers identified with particular rockets. Reporters and photographers spent hours on Cocoa Beach, or along the jetties at Port Canaveral, looking northward in hope of seeing a launch. Terse official announcements were made after each launch, since the roar of rocket motors could not be hushed. Usually the statement merely said that the launch was a success, a partial success or a failure.

Project Vanguard, which began in 1955, marked a change in the government's handling of information related to rocket systems. Vanguard's objective was to place a ball-shaped satellite into Earth orbit as the U.S. contribution to the International Geophysical Year, a worldwide study of atmospheric and space phenomena. This would be the first U.S. attempt to launch a satellite into orbit. Although the Naval Research Laboratory developed the carrier rocket and its satellite, the program was non-military. Detailed descriptions of the rocket, satellite and tracking network were released to the press regularly as work progressed.

The press corps was invited to enter Cape Canaveral and cover the first Vanguard launch attempt in December 1957. Reporters and radio and television commentators received briefings, and communications lines were installed for them. On Dec. 6 they were taken to view the launch from an elevated site on the Cape.

To their astonishment, Vanguard rose several feet, then settled back slowly and exploded in flames.

Despite the outcome of the launch, a precedent had been established for press coverage. Thereafter, the commander of the Atlantic Missile Range (later the Eastern Test Range) arranged for reporters to cover all but the classified military launches. In return, the resident press agreed not to publish information in advance, not to report on postponements, and to report their stories only "when fire appeared in the tail."

Manned space launches bring thousands of visitors to KSC, ranging from high-ranking officials, famous entertainers and industrialists, to members of space workers' families.

This arrangement governed coverage when the Army Ballistic Missile Agency prepared for its attempt to launch a satellite. A few guarded, brief items appeared in the newspapers as launch day neared, but little was known about either the Juno I rocket or the satellite. Canvas shrouds hung from the service structure at Complex 26 to conceal the vehicle's shape as long as possible. This rocket, a Jupiter C with an added fourth stage, roared into the night on Jan. 31, 1958, successfully carrying the first U.S. satellite, Explorer 1, into Earth orbit.

When NASA was organized in October 1958, it liberalized the press policy. Every detail of preparation for Project Mercury flights received saturation coverage by television, radio, films, and newspapers. Lt. Col. John Powers, assigned to NASA from the Air Force, attained national fame as the "Voice of Mercury." The media reported Gemini launches with even more extensive coverage.

Anticipating requirements for the Apollo program, KSC constructed a press site at Launch Complex 39. A sheltered grandstand accommodates 350 reporters. Television and radio networks built trailer studios or buildings next to the grandstand. The facilities were ready for Apollo 4 in November 1967.

Attendance by the media and contractor public relations personnel at a KSC launch ebbs and flows as the space program continues to evolve. The first Saturn V launch drew only 510 requests for accreditation. But media and contractor public relations representation swelled as the Apollo program progressed, and by the time of the Apollo 11 lunar landing in 1969, NASA's resources were being tested. Apollo 11 drew 3,497 requests for accreditation; 1,788 persons were actually on hand for the historic event.

After Apollo 11, press attendance slacked off. Apollo 17, the last in the program and the only night launch, prompted a resurgence in attendance figures, as did the Apollo-Soyuz Test Project in 1975.

The first Shuttle launch was the biggest show ever in the manned space program, with 2,707 accredited media and public relations representatives present. In September 1988 the flight of Discovery (STS-26), the first Shuttle mission after the Challenger accident, drew a record 5,200 requests for accreditation. The number of media and contractor public relations representatives who showed up for the launch was 2,468—the second highest ever for a launch from KSC.

Prelaunch briefings are arranged for the media, at which top NASA managers and officials from other organizations discuss preparations for the mission and its payloads. After the launch, press representatives are briefed on the results of that phase of the mission.

The Vehicle Assembly Building and two Shuttle launch pads form a backdrop for Spaceport USA, the KSC visitors center.

chapter 19

FACING the FUTURE

In July 1987, Kennedy Space Center celebrated 25 years as the nation's gateway to space. It was an occasion to reflect proudly on past achievements, but also to look ahead. KSC will continue to play a dominant role in realizing America's future in space as the United States heads toward the end of the millennium in the year 2000.

One of the foundations for that future will be Space Station Freedom. In January 1984 President Ronald Reagan directed NASA to develop a permanently manned space station. Canada, Japan and nine of the 13 members of the European Space Agency (ESA) have joined with the United States to design and develop this multipurpose facility, scheduled to be fully operational in the late 1990s.

When fully operational and permanently manned, Space Station Freedom will facilitate not only more detailed exploration of our own planet, but it will also serve as a steppingstone to the Moon and other planets.

"Let us keep in focus the concept that underlies the space station endeavor, providing a sense of urgency and direction," said James Odom, NASA associate administrator for space station, 1988-89. "The concept is leadership. The space station is about leadership in space. Not leadership in the 1980s or early 1990s, but leadership later on, in the late 1990s and well into the next century. The space station is about a time when NASA's recovery from the loss of Challenger is complete, when the Soviets will have had years of operational experience aboard the Mir space station, when the Hubble Space Telescope and the other Great Observatories will be in need of upgrading and refurbishment, when Europe will be close to its stated goal of independent capabilities in space, when the technical talents of Japan will have focused upon launch vehicles and spacecraft, and when the time for planning exploration beyond Earth orbit will have arrived. The space station is about the future."

Space Station Freedom will be a laboratory, an observatory, a servicing and repair facility, and a staging area for future manned and unmanned exploration. The station will have four pressurized modules affixed to a horizontal boom. Two of these are research laboratories provided by ESA and Japan. The United States will provide another, plus a fourth module that will be living quarters for a crew of four.

NASA plans a phased approach toward construction of the space station. Six Shuttle flights in the mid-1990s will be required to achieve a man-tended capability, where a Shuttle crew can remain in space 16 to 28 days with the orbiter attached to the station. After another 11 Shuttle flights, enough additional equipment will be delivered and assembled to achieve a permanently manned capability. Astronauts will be able to live and work in the station year-round, visited periodically by a Space Shuttle orbiter delivering supplies.

KSC will make a major contribution to the space station program. Space station hardware will be processed and checked out at the Space Station Processing Facility (SSPF) that NASA is building in the KSC Industrial Area, and Shuttles launched by the KSC team will deliver the elements of Freedom into space.

The SSPF is the largest single new construction project undertaken at the space center since the Apollo era. Scheduled for completion in 1994, the 457,000-square-foot facility is designed to make use of existing KSC systems whenever possible to keep costs down. The three-story facility will become a regular stop for KSC public tours, and its design features a viewing gallery for visitors.

In addition to the scientific activity which Space Station Freedom will allow, NASA also plans a number of other exploratory endeavors aimed at increasing human knowledge of the universe. The Great Observatories program was inaugurated with the launch into orbit of the Hubble Space Telescope (HST) and Gamma Ray Observatory (GRO). The Advanced X-Ray Astrophysics Facility (AXAF) and the Space Infrared Telescope Facility (SIRTF) will be launched later in the decade.

The Great Observatories program will yield a more complete picture of the world beyond Earth's atmosphere. Each Great Observatory will examine the universe through different bands of the electromagnetic spectrum. Visible light shows us sizes, shapes and textures. Other wavelengths, not visible to the naked eye, reveal chemical and physical forces at work: gamma rays, X-rays, ultraviolet, infrared, microwave and radio wavelengths. Each observatory will yield data on the types of bodies and phenomena which emanate a particular type of wavelength, such as the X-rays which black holes emit and which AXAF will detect. Like pieces

Workers prepare the Magellan spacecraft for testing. The first U.S. planetary explorer to be launched in nearly a decade, Magellan mapped 90 percent of the surface of Venus. White thermal blankets cloak most of the spacecraft's major parts to maintain temperature control.

from a puzzle, scientists will use the individual pictures of the universe, revealed through the Great Observatories, to fill in many of the gaps in our knowledge.

The Great Observatories will fulfill their missions from positions in Earth orbit. NASA also plans to send probes and orbiters into space that will come much closer to their physical targets. Starting with the Magellan mission launched in May 1989 to map the surface of Venus, three planetary explorers have been carried aloft by Space Shuttles. These low-cost spacecraft, utilizing spare parts and existing technology, will add to the foundation of knowledge accrued from earlier missions. Magellan was followed by Galileo in October 1989, on a long journey to Jupiter. Then Ulysses, formerly known as the International Solar Polar Mission, was launched in October 1990. It will reconnoiter the poles of the Sun and the space around them.

The Space Transportation System which will send these high-technology emissaries on their respective odysseys will itself continue to evolve. NASA has already initiated an acquisition plan for an advanced solid rocket motor (ASRM) to increase the Shuttle's lift capability by 12,000 pounds (5,443 kilograms). Incorporating state-of-the-art solid rocket motor design technology and extensive automated building techniques, the ASRM will be on hand in the mid-1990s to launch elements of the space station.

A project still on the drawing board is an unmanned version of the Shuttle. Known as Shuttle-C, this partially reusable cargo vehicle would capitalize on the existing Space Transportation System infrastructure at KSC and elsewhere to keep costs down. With its heavy lift capability—anywhere from 100,000 to 170,000 pounds (45,360 to 77,112 kilograms)—Shuttle-C could reduce by 50 percent the number of launches and length of assembly time for space station elements. It also could be used to carry into orbit scientific spacecraft. Among the candidate missions under study are two of the Great Observatories, AXAF and SIRTF. Finally, a cargo Shuttle could serve as a test bed for new Shuttle elements like the ASRM.

The Space Station, Great Observatories and interplanetary missions are only some of the activities NASA is planning in the decade ahead. Like the pioneering missions upon which they will build, they too are merely preludes to explorations still to come in the next century. The United States space program will always be open-ended, a continuing scientific exploration of the unknown, with benefits that are sometimes predictable, but very often completely unforeseen.

The Galileo planetary mission to Jupiter will send an instrumented probe into the Jovian atmosphere and a spacecraft to orbit the planet and collect data about its satellites. Galileo will build on knowledge gained through the Voyager expeditions; this collage of Jupiter and its four moons was assembled from Voyager-transmitted imagery.

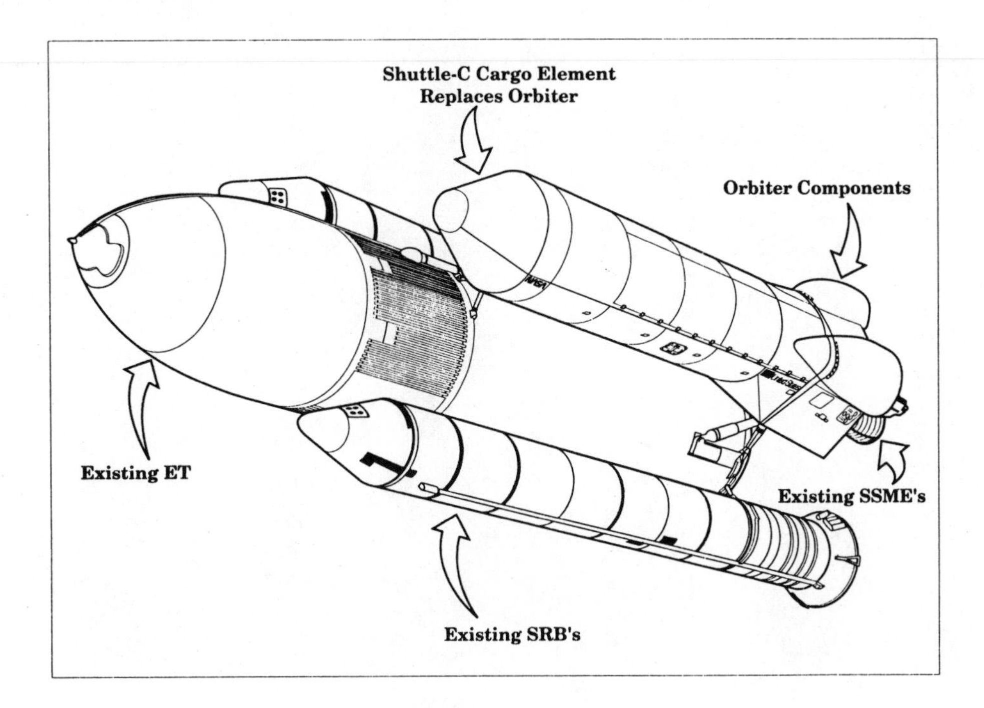

Shuttle-C, an unmanned cargo variant of the Shuttle, would draw heavily upon the existing Space Transportation System to allow near-term development while keeping costs and risk down.

"Some call space the endless frontier," James Beggs, NASA administrator from 1981 to 1986, said in an address to agency workers, "and it is, indeed, endless, because no matter how far you go, there is always further to go. This frontier offers countless new opportunities to exploit. These opportunities, too, are literally endless. And we have just begun to grasp them."

The nation's spaceport will play a major role in the expansion of humanity beyond its native planet. The story of Kennedy Space Center, too, is open-ended. The future lies waiting.